Ergebnisse der Mathematik und ihrer Grenzgebiete

W0049804

Band 48

Herausgegeben von
P. R. Halmos · P. J. Hilton · R. Remmert · B. Szőkefalvi-Nagy

Unter Mitwirkung von
L. V. Ahlfors · R. Baer · F. L. Bauer · R. Courant
A. Dold · J. L. Doob · S. Eilenberg · M. Kneser · G. H. Müller
M. M. Postnikov · B. Segre · E. Sperner

Geschäftsführender Herausgeber: P. J. Hilton

A. Ionescu Tulcea and C. Ionescu Tulcea

Topics in the Theory of Lifting

Springer-Verlag Berlin Heidelberg GmbH 1969

A. Ionescu Tulcea and C. Ionescu Tulcea

Northwestern University, Evanston, Ill.

ISBN 978-3-642-88509-9   ISBN 978-3-642-88507-5 (eBook)
DOI 10.1007/978-3-642-88507-5

All rights reserved. No part of this book may be translated or reproduced in any form without written permission from Springer-Verlag Berlin Heidelberg.

Library of Congress Catalog Card Number 69-18412.
Title No. 4592

© Springer-Verlag Berlin Heidelberg 1969
Originally published by Springer-Verlag Berlin · Heidelberg in 1969
Softcover reprint of the hardcover 1st edition 1969

Preface

The problem as to whether or not there exists a lifting of the M_R^∞ space[1]) corresponding to the real line and Lebesgue measure on it was first raised by A. Haar. It was solved in a paper published in 1931 [102] by J. von Neumann, who established the existence of a lifting in this case. In subsequent papers J. von Neumann and M. H. Stone [105], and later on J. Dieudonné [22], discussed various algebraic aspects and generalizations of the problem.

Attemps to solve the problem as to whether or not there exists a lifting for an arbitrary M_R^∞ space were unsuccessful for a long time, although the problem had significant connections with other branches of mathematics.

Finally, in a paper published in 1958 [88], D. Maharam established, by a delicate argument, that a lifting of M_R^∞ always exists (for an arbitrary space of σ-finite mass). D. Maharam proved first the existence of a lifting of the M_R^∞ space corresponding to a product $X=\prod_{i\in I}\{a_i,b_i\}$ and a product measure $\mu=\bigotimes_{i\in I}\mu_i$, with $\mu_i\{a_i\}=\mu_i\{b_i\}=\frac{1}{2}$ for all $i\in I$. Then, she reduced the general case to this one, via an isomorphism theorem concerning homogeneous measure algebras [87], [88].

A different and more direct proof of the existence of a lifting was subsequently given by the authors in [65]. A variant of this proof is presented in chapter 4.

It should be noticed that it is the "positivity" of the linear lifting (or correspondingly the "continuity property") which makes the proof of its existence difficult. However, it is precisely this property of the lifting which is important in applications.

The fact that there always exists a lifting (for strictly localizable spaces) has important applications. For instance: in the representation of integral operators, in the problem of disintegration of measures and ergodic theory, in the existence of separable modifications for stochastic processes and in the problem of point realizations of automorphisms

[1]) This notation is explained in section 7, chapter 1.

of L_R^∞ spaces[2]). Many classical theorems can be stated and proved in final form using the notion of lifting.

Most of the results given in this volume are due to the authors and were established in their papers quoted in the bibliography. The presentation makes use of various improvements in methods and results which were obtained subsequently by the authors.

Results concerning liftings commuting with sets of mappings (for instance the fact that if X is a locally compact group and μ a left invariant Haar measure on X, then there is always a lifting of the corresponding M_R^∞ space commuting with the left translation of X) are not included here. Complete details can be found in [54] and [72].

The volume is divided into ten main chapters. In chapter 1 we outline the setting for the theory of integration that is used in the book. The approach that we develop is based on the notion of upper integral. It provides a unified treatment for Bourbaki's integrals (both the usual and the essential integral) and for the integral in the setting of abstract measure spaces.

In chapter 2 we introduce the basic notion of admissible subalgebra of M_R^∞ and we study projections onto admissible subalgebras. Admissible subalgebras are used in the proof of the existence of a lifting.

In chapter 3 we introduce the notions of lifting, linear lifting and lower density (for an admissible subalgebra). We show in particular that the existence of a linear lifting is equivalent with the existence of a lifting.

The existence of a lifting (for strictly localizable spaces) is proved in chapter 4. The proof makes use of admissible subalgebras and an ergodic theorem concerning increasing sequences of projections (i. e. the appropriate version of the martingale convergence theorem, which is proved in Appendix I at the end of the volume). In view of later applications, we also show here how to define liftings for functions having as range a completely regular space.

In chapter 5 we define and study the topologies associated with lower densities and with liftings of an admissible subalgebra. Most of the results of this chapter were proved in [57].

In chapter 6 we discuss the integrability and measurability of functions with values in Banach spaces. The definitions and results are used in the next chapter.

In chapter 7 we prove a general integral representation theorem (without any separability assumptions). This theorem yields as corol-

---

[2]) For further results concerning liftings and differentiation of measures, the reader may consult the paper [78], which appeared after the manuscript of this volume was completed and sent for publication.

laries the Dunford-Pettis theorem and the Dunford-Pettis-Phillips theorem. The dual of L^p_E $(1 \leqslant p < \infty)$ is obtained without any separability hypotheses. This chapter contains also a proof of Strassen's integral representation theorem, again without any separability assumptions. Using liftings for functions having as range a completely regular space, we give a short and direct proof of the fact that certain stochastic processes admit separable modifications.

In chapter 8 we introduce and discuss the notion of strong lifting and almost strong lifting, in the setting of locally compact spaces and Radon measures. It is shown that if X is metrizable and μ is a Radon measure on X having X for support, then there is a strong lifting of the corresponding M^∞_R space. A series of examples in which X is not metrizable but a strong lifting still exists, are given (if X is a locally compact group and μ a left invariant Haar measure on X, then there exists a strong lifting). The chapter ends with a discussion of strong and almost strong liftings in the setting of topological (non locally compact) spaces. There is also an appendix on Borel liftings.

In chapter 9 we first prove a theorem about domination of measures, as a corollary to Strassen's theorem proved in chapter 7. From this theorem we obtain directly the disintegration of measures in the case of compact spaces and continuous mappings. The general disintegration theorem is proved in the last section of the chapter. We also show in this chapter that the existence of a strong lifting and that of a disintegration are in a certain sense equivalent problems. Among the notions introduced in this chapter in connection with the study of disintegration of measures we would like to mention that of "appropriate family".

In chapter 10 we show that every automorphism of an L^∞_R space is induced by a point mapping.

The volume ends with two appendices and a short list of open problems (the main one, at present, being the existence of a strong lifting).

The authors wish to thank Miss Vera Fisher for her excellent typing of the manuscript.

During the preparation of this volume the authors were supported by the U.S. Army Research Office (Durham, N. C.) under contract DA-31-124-ARO(D)-288. The authors wish to extend their thanks to Dr. A. S. Galbraith and the administrative staff of that office for their cooperation.

Evanston, March 1969

A. Ionescu Tulcea
C. Ionescu Tulcea

Contents

CHAPTER I

Measure and integration

In this chapter we outline the setting for the theory of measure and integration that will be used in this book. The approach that we develop is essentially based on the notion of upper integral. It has the advantage that it provides a unified treatment for Bourbaki's integrals (both the usual and the "essential" integral) and for the integral in abstract measure spaces. This chapter is self-contained. We give complete definitions and complete statements of the most important results in the theory. The proofs however are omitted in most cases.

1. The upper integral

Let X be a set and let $\bar{R}_+^X$ be the set of all mappings $f: X \to \bar{R}_+$ (here $\bar{R}$ is the extended real line and $\bar{R}_+ = \{x \in \bar{R} \mid x \geqslant 0\} = [0, +\infty]$).

Definition 1. – *A mapping $N: \bar{R}_+^X \to \bar{R}_+$ is an upper integral on X if it satisfies the following conditions:*

a) $N(0)=0$;

b) $N(\lambda f)=\lambda N(f)$ *for all* $f \in \bar{R}_+^X$ *and all* $\lambda > 0$;

c) *If* $f \leqslant \sum_{n=1}^{\infty} g_n$ *then* $N(f) \leqslant \sum_{n=1}^{\infty} N(g_n)$;

d) *If* (f_n) *is an increasing sequence then* $N(\sup_n f_n) = \sup_n N(f_n)$.

If N is an upper integral on X and if for each $1 \leqslant p < +\infty$ we define $N_p: \bar{R}_+^X \to \bar{R}_+$ by the equations

$$N_p(f) = (N(f^p))^{1/p}, \quad f \in \bar{R}_+^X$$

then N_p is an upper integral on X.

From this point, up to and including section 8, N will denote a fixed upper integral on X.

A function $f \in \bar{R}^X$ is called *N-negligible* if $N(|f|)=0$. A set $A \subset X$ is called *N-negligible* if φ_A is *N-negligible.*

When there is no ambiguity we shall say *negligible* instead of *N-negligible.*

We say that a property $P(x)$ holds *N-almost everywhere on a set* $A \subset X$ or equivalently holds *for almost every* $x \in A$ *with respect to* N if the set of all $x \in A$ for which $P(x)$ is not true is *N-negligible.*

When there is no ambiguity we shall say almost everywhere instead of N-almost everywhere and for almost every $x \in A$ instead of for almost every $x \in A$ with respect to N.

We shall now point out several properties of *N-negligible sets* and *N-negligible functions.*

(1) If A is negligible and $B \subset A$, then B is negligible.

(2) If (A_n) is a sequence of negligible sets then $\bigcup_n A_n$ and $\bigcap_n A_n$ are negligible.

(3) A function $f \in \bar{R}^X$ is negligible if and only if $S_f = \{x \mid f(x) \neq 0\}$ is negligible.

(4) If $f \in \bar{R}^X$ and $N(|f|) < +\infty$ then the set $\{x \mid |f(x)| = +\infty\}$ is negligible, i.e., f is finite almost everywhere.

Let $f: X \to \bar{R}$. For each $1 \leqslant p < +\infty$ we define $N_p(f)$ by

$$N_p(f) = N_p(|f|).^{1})$$

We denote by $\mathscr{F}^p(X,N)$ the set of all $f: X \to R$ for which

$$N_p(f) < +\infty.$$

The restriction of N_p to $\mathscr{F}^p(X,N)$ will be denoted by the same symbol.

When there is no ambiguity we shall write $\mathscr{F}^p$ instead of $\mathscr{F}^p(X,N)$.

Theorem 1. – *For each* $1 \leqslant p < +\infty$ *the set* $\mathscr{F}^p$ *is a vector space,* N_p *is a seminorm on* $\mathscr{F}^p$ *and* $\mathscr{F}^p$ *is complete with respect to this seminorm.*

From now on we shall always assume that $\mathscr{F}^p$ is endowed with the topology defined by the seminorm N_p; we shall call this topology the *topology of mean convergence of order p* (for $p=1$ we call it simply the *topology of mean convergence*) on $\mathscr{F}^p$. If $A \subset \mathscr{F}^p$ then we shall usually call topology of mean convergence of order p on A (or topology of mean convergence if $p=1$) the topology induced on A by the topology of $\mathscr{F}^p$.

---

[1]) If $g: X \to \bar{R}$ and $h: X \to \bar{R}$ coincide almost everywhere then $N_p(f) = N_p(g)$ for every $1 \leqslant p < +\infty$.

2. The spaces $\mathscr{L}^p$ and L^p $(1 \leqslant p < +\infty)$

Let now $\mathscr{R} \subset \mathscr{F}^1(X,N)$ be a set with the following properties:

(L_1) *$\mathscr{R}$ is a vector space;*
(L_2) *If $\varphi \in C_R(R)$, $\varphi(0)=0$ and $f \in \mathscr{R}$, then $\varphi \circ f \in \mathscr{R}$;*
(L_3) *$N(f+g)=N(f)+N(g)$ for $f \in \mathscr{R}_+$ and $g \in \mathscr{R}_+$.*[1])

Remark. It follows easily from conditions (L_1) and (L_2) that $\mathscr{R}$ is a Riesz space (for the usual pointwise order relation) and that $\mathscr{R}_+$ spans $\mathscr{R}$. From (L_2) we also deduce:

(C_1) $f \in \mathscr{R} \Rightarrow \inf(f,1) \in \mathscr{R}$;
(C_2) $f \in \mathscr{R}$ and $1 \leqslant p < +\infty \Rightarrow |f|^p \in \mathscr{R}$;
(C_3) $\mathscr{R}$ is an algebra (note that for $f \in \mathscr{R}$ and $g \in \mathscr{R}$ we have $4fg = (f+g)^2 - (f-g)^2$ and use (C_2) and (L_1));
(C_4) $\mathscr{R} \subset \mathscr{F}^p(X,N)$ for each $1 \leqslant p < +\infty$.

Definition 2. *Let $1 \leqslant p < +\infty$. We define $\mathscr{L}^p(X,N,\mathscr{R})$ to be the closure of $\mathscr{R}$ in $\mathscr{F}^p(X,N)$. We define $L^p(X,N,\mathscr{R})$ to be the separated space associated with $\mathscr{L}^p(X,N,\mathscr{R})$.*

Clearly $\mathscr{L}^p(X,N,\mathscr{R})$ and $L^p(X,N,\mathscr{R})$ are vector spaces. The *canonical mapping* of $\mathscr{L}^p(X,N,\mathscr{R})$ onto $L^p(X,N,\mathscr{R})$ will be denoted by $f \to \tilde{f}$. Note that if f and g belong to $\mathscr{L}^p(X,N,\mathscr{R})$ then $\tilde{g} = \tilde{f}$ if and only if the set $\{x \mid g(x) \neq f(x)\}$ is N-negligible.

Recall that the norm on $L^p(X,N,\mathscr{R})$ (which will also be denoted by N_p) is defined by

$$N_p(\tilde{f}) = N_p(f)$$

for each $\tilde{f} \in L^p(X,N,\mathscr{R})$.

When there is no ambiguity we shall write $\mathscr{L}^p$ instead of $\mathscr{L}^p(X,N,\mathscr{R})$ and L^p instead of $L^p(X,N,\mathscr{R})$.

Remarks. – 1) We want to stress the fact that the definition of the space $\mathscr{L}^p(X,N,\mathscr{R})$ *depends essentially on the vector space* $\mathscr{R}$, while the definition of the spaces $\mathscr{F}^p(X,N)$ *depends only on the upper integral.*

2) A function $f: X \to R$ belongs to $\mathscr{L}^p(X,N,\mathscr{R})$ if and only if for each $\varepsilon > 0$ there is $f_\varepsilon \in \mathscr{R}$ such that $N_p(f - f_\varepsilon) < \varepsilon$.

3) The *completeness* of $\mathscr{L}^p(X,N,\mathscr{R})$ and $L^p(X,N,\mathscr{R})$ is a consequence of their definition and the completeness of $\mathscr{F}^p(X,\mathrm{N})$. Therefore $L^p(X,N,\mathscr{R})$ $(1 \leqslant p < +\infty)$ is a *Banach space* when endowed with the norm N_p.

---

[1]) For any ordered vector space E and set $S \subset E$ we write $S_+ = \{x \in S \mid x \geqslant 0\}$.

4) If $f \in \mathcal{F}^p(X,N)$ coincides almost everywhere with a function belonging to $\mathcal{L}^p(X,N,\mathcal{R})$ then $f \in \mathcal{L}^p(X,N,\mathcal{R})$. In particular $f \in \mathcal{L}^p(X,N,\mathcal{R})$ if $f: X \to R$ is N-negligible.

Let $A \subset X$, $f: A \to \bar{R}$. We say that f is *defined almost everywhere* if $\complement A$ is negligible. If $g: X \to \bar{R}$, we say that g is *equivalent* with f if $g(x)=f(x)$ almost everywhere.

A function f with values in $\bar{R}$, defined almost everywhere, is said to be *p-integrable* $(1 \leqslant p < +\infty)$ *with respect to* $(N,\mathcal{R})$ if it is equivalent with a function $g \in \mathcal{L}^p(X,N,\mathcal{R})$. If $p=1$ then instead of 1-integrable with respect to $(N,\mathcal{R})$ we shall usually say *integrable with respect to* $(N,\mathcal{R})$ or $(N,\mathcal{R})$*-integrable.*

A function f belongs therefore to $\mathcal{L}^p(X,N,\mathcal{R})$ $(1 \leqslant p < +\infty)$ if and only if it is p-integrable with respect to $(N,\mathcal{R})$, has for *domain* X and *takes values in* R.

If f is a function with values in $\bar{R}$, defined almost everywhere, we define $N_p(f)=N_p(g)$ $(1 \leqslant p < +\infty)$ if $g: X \to \bar{R}$ and g is equivalent with f. If f is p-integrable with respect to $(N,\mathcal{R})$ and if $g \in \mathcal{L}^p(X,N,\mathcal{R})$ is equivalent with f, then we shall sometimes write $\tilde{f}=\tilde{g}$. We define mean convergence of order p in the natural way for functions that are p-integrable with respect to $(N,\mathcal{R})$.

When there is no ambiguity we shall say p-integrable instead of p-integrable with respect to $(N,\mathcal{R})$ and integrable instead of $(N,\mathcal{R})$-integrable.

From this point, up to and including section 8, $\mathcal{R} \subset \mathcal{F}^1(X,N)$ *will be a fixed set with the properties* (L_1), (L_2), (L_3).

Several properties of the spaces $\mathcal{L}^p$ and $L^p (1 \leqslant p < +\infty)$ are given in the theorems below:

Theorem 2. – *Let $\mathcal{E} \subset \mathcal{L}^p$ be a set dense in $\mathcal{L}^p$. Then for every $f \in \mathcal{L}^p$ there exists a sequence (f_n) of functions belonging to $\mathcal{E}$ and having the following properties:*

2.1) *The sequence (f_n) converges to f in mean of order p;*

2.2) *The sequence $(f_n(x))$ converges to $f(x)$ almost everywhere;*

2.3) *There is $g: X \to \bar{R}_+$ with $N_p(g) < +\infty$ such that $|f_n| \leqslant g$ for each n.*

In particular the result in Theorem 2 is valid for $\mathcal{E}=\mathcal{R}$. Moreover if $\mathcal{E}=\mathcal{R}$ and if $|f(x)| \leqslant M$ for all $x \in X$, then we may suppose that $|f_n(x)| \leqslant M$ for all $x \in X$ and all n.

The next result shows, in particular, that $\mathcal{L}^p$ is a Riesz space.

Theorem 3. – *For each $f \in \mathcal{L}^p$, the function $|f|$ belongs to $\mathcal{L}^p$ and the mapping $f \to |f|$ of $\mathcal{L}^p$ into $\mathcal{L}^p$ is uniformly continuous. If $f \in \mathcal{L}^p$ and $g \in \mathcal{L}^p$ then* $\sup(f,g) \in \mathcal{L}^p$ *and* $\inf(f,g) \in \mathcal{L}^p$.

For $\tilde{f} \in L^p$ and $\tilde{g} \in L^p$ we write

$$\tilde{f} \leqslant \tilde{g} \Leftrightarrow f(x) \leqslant g(x) \text{ almost everywhere.}$$

It is clear that this is an order relation in L^p and that L^p is a Riesz space with respect to this order relation. Moreover the Riesz space L^p is completely reticulated as follows from:

Theorem 4. – *Let $H \subset L^p_+$ be a set directed for the relation $\leqslant$ and such that $\sup_{\tilde{h} \in H} N_p(\tilde{h}) < \infty$. Then H has a supremum $\tilde{h}_\infty$ in L^p and*

$$\lim_{\tilde{h} \in H} N_p(\tilde{h}_\infty - \tilde{h}) = 0.$$

For any increasing sequence $(\tilde{h}_n)$ of elements in H such that $\sup_n N_p(\tilde{h}_n) = \sup_{\tilde{h} \in H} N_p(\tilde{h})$ we have

$$\sup_n \tilde{h}_n = \tilde{h}_\infty \quad \textit{and} \quad \lim_n N_p(\tilde{h}_\infty - \tilde{h}_n) = 0.$$

It is clear that there exists an increasing sequence $(\tilde{h}_n)$ of elements in H such that $\sup_n N_p(\tilde{h}_n) = \sup_{\tilde{h} \in H} N_p(\tilde{h})$.

Theorem 5 (Monotone convergence). – *Let (f_n) be an increasing sequence of functions belonging to $\mathscr{L}^p_+$; then the following two assertions are equivalent:*

i) *The pointwise supremum f of (f_n) is p-integrable;*
ii) $\sup_n N_p(f_n) < +\infty$.

If ii) *is satisfied then we have*

$$\lim_n N_p(f - f_n) = 0.$$

A very useful result is the following:

Theorem 6 (Lebesgue). – *Let (f_n) be a sequence of functions belonging to $\mathscr{L}^p$ with the following properties:*

1) *The sequence $(f_n(x))$ converges almost everywhere to a limit $f(x) \in R$;*

2) *There is $g: X \to \bar{R}_+$ with $N_p(g) < +\infty$ such that for each n, $|f_n(x)| \leqslant g(x)$ almost everywhere.*

Then the function f (defined almost everywhere) is p-integrable and the sequence (f_n) converges to f in mean of order p.

A useful characterization of $\mathscr{L}^p$ spaces is the following:

Theorem 7. – *A function* $f: X \to R$ *belongs to* $\mathscr{L}^p$ *if and only if* $|f|^{p-1} f \in \mathscr{L}^1$.

Let now $1 \leqslant p < +\infty$ and $A \subset X$. It follows from Theorem 7 that

$$\varphi_A \in \mathscr{L}^1 \Leftrightarrow \varphi_A \in \mathscr{L}^p.$$

We shall now close this section with the following result which is an immediate consequence of Theorems 2 and 6:

Theorem 8. – *Let* $1 \leqslant p < +\infty$ *and let* $\varphi \in \mathscr{L}^p$ *and* $f \in \mathscr{L}^p$. *Suppose* φ *bounded. Then* $\varphi f \in \mathscr{L}^p$.

From Theorem 8 it follows in particular that if $A \subset X$ is a set such that $\varphi_A \in \mathscr{L}^1$ and if $f \in \mathscr{L}^p$ then $\varphi_A f \in \mathscr{L}^p$.

3. The integral

In the previous section we defined the notion of integrable function. We shall now define the integral of an integrable function.

Definition 3. – *We call integral associated with* $(N, \mathscr{R})$ *the unique linear mapping* $\mu_{(N,\mathscr{R})}: \mathscr{R} \to R$ *satisfying*

$$\mu_{(N,\mathscr{R})}(f) = N(f) \quad \textit{for} \quad f \in \mathscr{R}_+.$$

When there is no ambiguity we write μ instead of $\mu_{(N,\mathscr{R})}$.

In other words, μ is the linear extension of $N|\mathscr{R}_+$ to the vector space $\mathscr{R}$ spanned by the cone $\mathscr{R}_+$ (see section 2, axiom (L_3) and the remark following it).

We remark that for each $f \in \mathscr{R}$ we have

$$|\mu(f)| \leqslant N(f).$$

Hence $\mu: \mathscr{R} \to R$ is a continuous linear mapping if $\mathscr{R}$ is endowed with the topology of mean convergence. It follows that there exists a *unique continuous linear extension* of μ to $\mathscr{L}^1$. The value of this linear functional for an element $f \in \mathscr{L}^1$ is denoted by $\int_X f\, d\mu$ (or $\int_X f\, d\mu_{(N,\mathscr{R})}$)[1]) or $\int f\, d\mu$ and is called *the integral of* f. Although the main properties of the integral are immediate consequences of the basic properties of the upper integral we shall point out the following:

---

[1]) When necessary we write

$$\int_X f(x)\, d\mu(x) \quad \text{or} \quad \int_X f(x)\, d\mu_{(N,\mathscr{R})}(x).$$

(1) For $f \in \mathscr{L}_+^1$, $\int\limits_X f\,d\mu = N(f)$.

(2) For each $g \in \mathscr{L}^1$, $\left|\int\limits_X g\,d\mu\right| \leqslant \int\limits_X |g|\,d\mu$.

(3) If $f \in \mathscr{L}^1, g \in \mathscr{L}^1$ are equivalent, then $\int\limits_X f\,d\mu = \int\limits_X g\,d\mu$.

When $A \subset X$ is such that $\varphi_A \in \mathscr{L}^1$ we shall sometimes write $\mu(A)$ instead of $\int\limits_X \varphi_A\,d\mu$.

4. Measurable functions

For each $f \in \bar{R}_+^X$ we define

$$\mathscr{D}_f = \{g \text{ integrable} \mid g \geqslant f\}.$$

Definition 4. – *We say that the upper integral N is regular if, for every* $f \in \bar{R}_+^X$,

$$N(f) = \begin{cases} \inf\{N(g) \mid g \in \mathscr{D}_f\} & \text{if } \mathscr{D}_f \neq \varnothing, \\ +\infty \quad \text{if } \mathscr{D}_f = \varnothing. \end{cases}$$

From this point on we always assume that the upper integrals we consider are regular.

We shall now introduce two important definitions.

Definition 5. – *A set $A \subset X$ is $(N, \mathscr{R})$-integrable if $\varphi_A \in \mathscr{L}^1(X, N, \mathscr{R})$. We denote by $\mathscr{B}_0(X, N, \mathscr{R})$ the set of all $(N, \mathscr{R})$-integrable sets $A \subset X$.*

Remarks. – 1) If A and B belong to $\mathscr{B}_0(X, N, \mathscr{R})$ then $A \cup B$, $A \cap B$, and $A - B$ belong to $\mathscr{B}_0(X, N, \mathscr{R})$.

2) If $A \subset X$ is such that $N(\varphi_A) < \infty$ then there exists $B \in \mathscr{B}_0(X, N, \mathscr{R})$ such that $B \supset A$.

In fact, let $g \in \mathscr{D}_{\varphi_A}$; then $N(g) < \infty$. Let $g_1 = \inf(1, g)$. Then $\varphi_A \leqslant g_1^n \leqslant g_1$ and $g_1^n \in \mathscr{L}^1$ for all n (see Theorem 8). The sequence (g_1^n) converges pointwise to a characteristic function φ_B which (by Lebesgue's theorem) belongs to $\mathscr{L}^1$.

When there is no ambiguity we shall say integrable instead of $(N, \mathscr{R})$-integrable and we shall write $\mathscr{B}_0$ instead of $\mathscr{B}_0(X, N, \mathscr{R})$.

Definition 6. – *A function $f: X \to \bar{R}$ is called $(N, \mathscr{R})$-measurable if given any $B \in \mathscr{B}_0(X, N, \mathscr{R})$ there is a sequence (h_n) of functions belonging to $\mathscr{R}$ such that*

$$\lim_n h_n(x) = f(x)$$

N-almost everywhere on B. We denote by $\mathscr{L}(X, N, \mathscr{R})$ the set of all $(N, \mathscr{R})$-measurable functions on X to R.

From the definition it follows immediately that $\mathscr{L}(X,N,\mathscr{R})$ is an *algebra* over R and a *Riesz space* (for the usual pointwise order).

When there is no ambiguity we shall say measurable instead of $(N,\mathscr{R})$-measurable and we shall write $\mathscr{L}$ instead of $\mathscr{L}(X,N,\mathscr{R})$.

A set $A \subset X$ is said to be $(N,\mathscr{R})$-*measurable* if φ_A is $(N,\mathscr{R})$-measurable. We denote by $\mathscr{B}(X,N,\mathscr{R})$ the set of all $(N,\mathscr{R})$-measurable sets $A \subset X$.

Proposition 1. – *A set* $A \subset X$ *belongs to* $\mathscr{B}(X,N,\mathscr{R})$ *if and only if* $A \cap B \in \mathscr{B}_0(X,N,\mathscr{R})$ *for every* $B \in \mathscr{B}_0(X,N,\mathscr{R})$.

When there is no ambiguity we shall say (as in the case of functions) measurable instead of $(N,\mathscr{R})$-measurable and we shall write $\mathscr{B}$ instead of $\mathscr{B}(X,N,\mathscr{R})$.

Proposition 2. – *The set* $\mathscr{B}$ *is a tribe* ($=\sigma$-*algebra*).

Let $A \subset X$ be $(N,\mathscr{R})$-measurable and $f: A \to \bar{R}$. We say that f is $(N,\mathscr{R})$-measurable if the mapping $f': X \to \bar{R}$ defined by

$$f'(x) = \begin{cases} f(x) & \text{if } x \in A, \\ 0 & \text{if } x \notin A \end{cases}$$

is $(N,\mathscr{R})$-measurable.

Remarks. – 1) Let f and g be two functions on X to $\bar{R}$ which coincide almost everywhere on each integrable set. If f is measurable then g is measurable.

2) Every p-integrable function $(1 \leqslant p < +\infty)$ is $(N,\mathscr{R})$-measurable (use Theorem 2).

3) If $f: X \to \bar{R}$ is measurable and $K \subset \bar{R}$ is a Borel set, then $f^{-1}(K)$ is measurable.

To prove 3) we reason as follows. We suppose first $f \in \mathscr{R}$. Assume that $K \subset \bar{R}$ is closed and such that $K \not\ni 0$. There is then a sequence (φ_n) of continuous functions on $\bar{R}$ to R which converges pointwise to φ_K; moreover we may suppose $\varphi_n(0)=0$ for all n. Then $(\varphi_n \circ f)$ converges pointwise to $\varphi_K \circ f = \varphi_{f^{-1}(K)}$. Since $\varphi_n \circ f \in \mathscr{R}$ for all n, we deduce that $\varphi_{f^{-1}(K)}$ is measurable. It follows then easily that $f^{-1}(K)$ is measurable for every Borel set $K \subset \bar{R}$.

Let now $f: X \to \bar{R}$ be measurable and let $K \subset \bar{R}$ be closed. We shall show that $f^{-1}(K)$ is measurable. For this it is enough to show that $B \cap f^{-1}(K)$ is measurable for every integrable set B. Let then $B \in \mathscr{B}_0(X,N,\mathscr{R})$ and let (h_n) be a sequence of functions in $\mathscr{R}$ which converges to $f(x)$ for every $x \in B_0$, where $B_0 \subset B$ and $B - B_0$ is negligible. If (U_n) is a sequence of open sets containing K and such that $\bigcap_n U_n = K$, then we have

$$B_0 \cap \Big(\bigcap_{p \geqslant 1} \bigcup_{m \geqslant 1} \bigcap_{n \geqslant m} h_n^{-1}(U_p) \Big) = B_0 \cap f^{-1}(K).$$

We conclude that $B \cap f^{-1}(K)$ is measurable. Since B was arbitrary, $f^{-1}(K)$ is measurable. It follows then easily that $f^{-1}(K)$ is measurable for every Borel set $K \subset \bar{R}$.

4) If $\varphi:\bar{R}\to\bar{R}$ is continuous, $\varphi(0)=0$ and if $f:X\to\bar{R}$ is measurable then $\varphi \circ f$ is measurable. In particular $|f|$ is measurable if $f:X\to\bar{R}$ is measurable.

A very useful result is the following:

Theorem 9. – *Let* $1\leqslant p<+\infty$ *and let* $f:X\to\bar{R}$. *The following assertions are equivalent:*

i) *The function* f *is p-integrable.*
ii) *The function* f *is measurable and* $N_p(f)<+\infty$.

The implication i)⇒ii) is immediate. We note then that it is enough to prove ii)⇒i) when f is bounded and vanishes outside an integrable set K (use Remark2) following Definition 5 and Lebesgue's theorem). By Definition 6 there exists a sequence (f_n) in $\mathcal{R}$ which converges almost everywhere on K to f. Let

$$\lambda > \sup_{x\in X} |f(x)|$$

and for each n let

$$K(n)=\{t|t\in K, |f_p(t)|<\lambda \quad \text{for} \quad p\geqslant n\}.$$

Then $\varphi_{K(n)}f_n$ is integrable for all n (use Theorem 8) and $(\varphi_{K(n)}f_n)$ converges almost everywhere (on X) to f. We may now use Lebesgue's theorem and conclude that f is p-integrable.

Let Y be a *metric space*, β the distance on Y, (f_n) a sequence of functions on a set A to Y and f a function on A to Y. We say that (f_n) converges to f, *uniformly with respect to* β, if for every $\varepsilon>0$ there is n_ε such that

$$\beta(f_n(x),f(x))\leqslant\varepsilon$$

for all $x\in A$, whenever $n\geqslant n_\varepsilon$.

On R we consider the usual distance d defined by

$$d(x,y)=|x-y|$$

for $x\in R$ and $y\in R$. On $\bar{R}$ we consider the distance δ "defined by"

$$\delta(x,y)=\left|\frac{x}{1+|x|}-\frac{y}{1+|y|}\right|$$

for $x\in\bar{R}$ and $y\in\bar{R}$.

From Remark 4) preceding Theorem 9 it follows that if u and v are measurable functions on X to R then the function

$$d(u,v):x \to d(u(x),v(x))$$

on X to R is measurable. It also follows that if u and v are measurable functions on X to $\bar{R}$ then the function

$$\delta(u,v):x \to \delta(u(x),v(x))$$

on X to R is measurable.

Theorem 10 (Egorov). – *Let (f_n) be a sequence of $(N,\mathcal{R})$-measurable functions on X to $\bar{R}$ and let $f: X \to \bar{R}$. Then the following assertions are equivalent:*

i) *For every set $B \in \mathcal{B}_0$, the sequence $(f_n(x))$ converges to $f(x)$ almost everywhere on B.*

ii) *For every $B \in \mathcal{B}_0$ and $\varepsilon > 0$ there exists $B_\varepsilon \in \mathcal{B}_0$, $B_\varepsilon \subset B$ such that $\mu(B - B_\varepsilon) \leqslant \varepsilon$ and such that $(f_n | B_\varepsilon)$ converges to $f | B_\varepsilon$ uniformly with respect to δ.*

Moreover if i) $(\equiv$ii)) *holds, then f is measurable.*

Remark. If the functions in the sequence (f_n) and f take values in R, then Theorem 10 remains valid if in ii) we replace

"with respect to δ" by "with respect to d".

The implication ii)$\Rightarrow$i) is immediate. To prove i)$\Rightarrow$ii) we reason as follows: Let $B \in \mathcal{B}_0$ and for each integer $p \geqslant 1$ and each integer $m \geqslant 1$ let

$$A_p^{(m)} = B \cap \bigcap_{i,j \geqslant p} \{x | \delta(f_i(x), f_j(x)) \leqslant 1/m\};$$

then $A_p^{(m)} \in \mathcal{B}_0$. The set C of all $x \in B$ such that $(f_n(x))$ converges to a limit can be written as

$$C = \bigcap_{m \geqslant 1} \bigcup_{p \geqslant 1} A_p^{(m)}.$$

We have for each $m \geqslant 1$

$$\mu(B) = \mu(C) = \lim_p \mu(A_p^{(m)}).$$

Let $\varepsilon > 0$. For each $m \geqslant 1$ there is $p(m) \geqslant 1$ such that

$$\mu(B - A_{p(m)}^{(m)}) \leqslant \varepsilon/2^m.$$

Define $B'_\varepsilon = \bigcap_m A_{p(m)}^{(m)}$; then $B'_\varepsilon \in \mathcal{B}_0$, $B'_\varepsilon \subset B$ and

$$\mu(B - B'_\varepsilon) \leqslant \sum_m \mu(B - A_{p(m)}^{(m)}) \leqslant \sum_m \varepsilon/2^m \leqslant \varepsilon.$$

It is clear that $(\delta(f_i, f_j) | B'_\varepsilon)$ converges uniformly to zero. If B_ε is the set of all $x \in B'_\varepsilon$ such that $(f_n(x))$ converges to $f(x)$, then obviously $(\delta(f_i, f_j) | B_\varepsilon)$ converges uniformly to zero. Hence $(f_n | B_\varepsilon)$ converges uniformly to $f | B_\varepsilon$ and hence ii) holds.

The fact that f is measurable follows immediately from the equivalence between i) and ii).

A function $f: X \to R$ is said to be *simple* if $f(X)$ is a finite set. It is clear that a simple function f *is integrable if and only if the set* $\{x | f(x) = a\}$ *is integrable for each* $a \in f(X)$, $a \neq 0$. We denote by $\mathcal{S}(X,N,\mathcal{R})$ the vector space of all *simple integrable functions.* It is clear that

$$\mathcal{S}(X,N,\mathcal{R}) \subset \mathcal{L}(X,N,\mathcal{R}).$$

When there is no ambiguity we shall write $\mathcal{S}$ instead of $\mathcal{S}(X,N,\mathcal{R})$.

Theorem 11. – *A function $f:X\to\bar{R}$ is measurable if and only if given any integrable set $K\subset X$ there is a sequence (s_n) of functions belonging to $\mathscr{S}$ such that $(s_n(x))$ converges to $f(x)$ almost everywhere on K.*

From Theorem 11 we deduce the following:

Corollary. – *The set $\mathscr{S}$ is dense in $\mathscr{L}^p$ for all $1\leqslant p<+\infty$.*

Let now n be an integer $\geqslant 1$ and consider

$$\bar{R}^n=\prod_{j=1}^{n}\bar{R}_j \quad (\bar{R}_j=\bar{R} \quad \text{for} \quad 1\leqslant j\leqslant n)$$

endowed with the product topology.

Theorem 12. – *Let $\varphi:\bar{R}^n\to\bar{R}$ be Borel measurable. For each $1\leqslant j\leqslant n$, let $f_j:X\to\bar{R}$ be a measurable mapping. Then $\varphi(f_1,\dots,f_n)$ is a measurable mapping.*

The next result establishes a relation between the $\mathscr{L}^p$ spaces and the seminorms N_p $(1<p<\infty)$:

Theorem 13. – *Let $1<p<\infty$ and $1<q<\infty$ be such that $1/p+1/q=1$. We have:*

13.1) *(Hölder's inequality) Let $f\in\mathscr{L}^p$ and $g\in\mathscr{L}^q$. Then $fg\in\mathscr{L}^1$ and*

$$N_1(fg)\leqslant N_p(f)N_q(g).$$

13.2) *Let $g\in\mathscr{L}^q$. Then*

$$N_q(g)=\sup\{|\textstyle\int fg\,d\mu|\mid f\in\mathscr{L}^p \quad \text{and} \quad N_p(f)\leqslant 1\}.$$

It can be shown (by the usual method) that for each $1<r<+\infty$, the space L^r is *uniformly convex* and hence *reflexive*.

Let $1<p<+\infty$ and $1<q<+\infty$ be such that $1/p+1/q=1$. For every $\tilde{g}\in L^q$ define the mapping $x'_{\tilde{g}}:L^p\to R$ by (use Theorem 13)

$$x'_{\tilde{g}}(\tilde{f})=\int fg\,d\mu \quad \text{for} \quad \tilde{f}\in L^p.$$

By Theorem 13, $x'_{\tilde{g}}\in(L^p)'$ and $\tilde{g}\to x'_{\tilde{g}}$ is a linear isometry of L^q into $(L^p)'$. Hence we may identify L^q with a closed linear subspace of $(L^p)'$. Since L^q separates the points of L^p we deduce that L^q is dense in $(L^p)'$ for the topology $\sigma((L^p)',L^p)$; since L^q is closed we obtain $L^q=(L^p)'$ (use the reflexivity of L^p).

Hence *the dual of* $L^p(1<p<+\infty)$ *is* $L^q(1<q<+\infty,\ 1/p+1/q=1)$.

The dual of L^1 will be discussed in section 8. We note that the above characterization of the dual of $L^p(1<p<+\infty)$ is obtained without any supplementary condition on $(X,N,\mathscr{R})$.

5. Further definitions and properties of measurable functions and sets

For $f:X\to\bar{R}$ and $g:X\to\bar{R}$ we write $f\equiv g$ whenever the set

$$\{x|f(x)\neq g(x)\}$$

is N-negligible (see also section 2).

Let now $\mathcal{N}(X,N)$ be the set of all N-negligible functions on X to R. Clearly $\mathcal{N}(X,N)$ is an ideal of $\mathcal{L}(X,N,\mathcal{R})$. Note that for f and g in $\mathcal{L}(X,N,\mathcal{R})$ we have $f\equiv g$ if and only if $f-g$ belongs to $\mathcal{N}(X,N)$.

We shall write

$$L(X,N,\mathcal{R})=\mathcal{L}(X,N,\mathcal{R})/\mathcal{N}(X,N)$$

and we shall denote by $\pi:f\to\tilde{f}$ the canonical mapping of $\mathcal{L}(X,N,\mathcal{R})$ onto the quotient space $L(X,N,\mathcal{R})$. For $\tilde{f}\in L(X,N,\mathcal{R})$ and $\tilde{g}\in L(X,N,\mathcal{R})$ we write

$$\tilde{f}\leqslant\tilde{g} \Leftrightarrow f(x)\leqslant g(x) \quad \text{almost everywhere.}$$

Clearly $L(X,N,\mathcal{R})$ is *an algebra as well as a Riesz space* (with respect to the previously defined order relation). The mapping π is a representation of the algebra $\mathcal{L}(X,N,\mathcal{R})$ onto the algebra $L(X,N,\mathcal{R})$.

When there is no ambiguity we shall write L instead of $L(X,N,\mathcal{R})$.

It is clear that the canonical mapping of $\mathcal{L}^p$ onto $L^p(1\leqslant p<+\infty)$ introduced in section 2, after Definition 2 can be identified with the restriction to $\mathcal{L}^p$ of the canonical mapping of $\mathcal{L}$ onto L. Also L^p can be *identified* with the image of $\mathcal{L}^p$ under the canonical mapping of $\mathcal{L}$ onto L. This justifies the use of the same notation for the "canonical mappings" of $\mathcal{L}^p$ onto L^p and $\mathcal{L}$ onto L. Similarly the order relation on L^p appears as the "restriction" of the order relation on L.

We denote by $\mathcal{N}_0(X,N)$ the set of all N-negligible parts of X.

When there is no ambiguity we write $\mathcal{N}_0$ instead of $\mathcal{N}_0(X,N)$.

For $A\in\mathcal{B}$, $B\in\mathcal{B}$ we write $A\,\triangle B=(A-B)\cup(B-A)$. We define an equivalence relation "$\equiv$" on $\mathcal{B}$ by writing

$$A\equiv B \Leftrightarrow A\,\triangle B\in\mathcal{N}_0.$$

We note that $A\equiv B$ if and only if $\varphi_A\equiv\varphi_B$.

We denote by $\mathcal{B}/\mathcal{N}_0$ the quotient space corresponding to this equivalence relation and by $B\to\tilde{B}$ the canonical mapping of $\mathcal{B}$ onto $\mathcal{B}/\mathcal{N}_0$.

We define an order relation on $\mathcal{B}/\mathcal{N}_0$ by writing for $\tilde{E}$ and $\tilde{F}$ in $\mathcal{B}/\mathcal{N}_0$

$$\tilde{E}\leqslant\tilde{F} \Leftrightarrow \tilde{\varphi}_E\leqslant\tilde{\varphi}_F.$$

The set $\mathscr{B}/\mathscr{N}_0$ endowed with this order relation is a σ-complete Boolean algebra.

Theorem 14. – *Let* $\mathscr{C} \subset \mathscr{B}$ *be such that* $\sup\{\tilde{K} | K \in \mathscr{C}\} = \tilde{X}$. *Then for every* $f \in \mathscr{L}_+$ *we have*

(1) $$\tilde{f} = \sup\{\tilde{\varphi}_K \tilde{f} | K \in \mathscr{C}\}.$$

In particular if $A \in \mathscr{B}$ *then*

(2) $$\tilde{A} = \sup\{\tilde{K} \cap \tilde{A} | K \in \mathscr{C}\}.$$

To prove (1) we note first that the case of an arbitrary $f \in \mathscr{L}_+$ can be reduced to the case when $0 \leqslant f \leqslant 1$. Let now $g \in \mathscr{L}_+$ be such that

$$0 \leqslant \tilde{\varphi}_K \tilde{f} \leqslant \tilde{g}$$

for all $K \in \mathscr{C}$. Then

$$\tilde{\varphi}_K = \tilde{\varphi}_K \tilde{f} + \tilde{\varphi}_K(\tilde{1} - \tilde{f}) \leqslant \tilde{g} + (\tilde{1} - \tilde{f}) = \tilde{1} + (\tilde{g} - \tilde{f})$$

for all $K \in \mathscr{C}$. Hence $\tilde{1} \leqslant \tilde{1} + (\tilde{g} - \tilde{f})$ and hence $\tilde{g} - \tilde{f} \geqslant 0$; we conclude $\tilde{g} \geqslant \tilde{f}$ and therefore (1) is proved. Clearly (2) is a consequence of (1).

Suppose now that $\mathscr{C} \subset \mathscr{B}$ is such that

$$\sup\{\tilde{K} | K \in \mathscr{C}\} = \tilde{X}.$$

We deduce then (use Theorems 4 and 14) that for every $B \in \mathscr{B}_0$ there is a countable set $\mathscr{C}_B \subset \mathscr{C}$ such that

$$B - \bigcup_{K \in \mathscr{C}_B} K$$

is negligible. On the basis of this remark we may prove the following:

Theorem 15. – *Let* $\mathscr{C} \subset \mathscr{B}$ *be such that* $\sup\{\tilde{K} | K \in \mathscr{C}\} = \tilde{X}$ *and let* $f: X \to \bar{R}$. *Then:*

15.1) *The function* f *is measurable if and only if for each* $K \in \mathscr{C}$, *the function* $\varphi_K f$ *is measurable.*

15.2) *We have* $f \equiv 0$ *if and only if* $\varphi_K f \equiv 0$ *for each* $K \in \mathscr{C}$.

6. Carathéodory measure

If for each $E \subset X$ we define $\mu_N^*(E)$ by

$$\mu_N^*(E) = N(\varphi_E)$$

then μ_N^* is an *outer measure in the sense of Carathéodory* ([50], [121]) called the outer measure associated with the upper integral N. Let us recall that a set $A \subset X$ is μ_N^**-measurable* if and only if it satisfies the following condition:

(*) For every $Y \subset X$, $\mu_N^*(Y) = \mu_N^*(Y \cap A) + \mu_N^*(Y \cap \complement A)$.

Let us also recall that the collection of all $A \subset X$ which are μ_N^*-measurable is a tribe (= σ-algebra) of subsets of X.

When there is no ambiguity we write μ^* instead of μ_N^*.

A function $f: X \to \bar{R}$ is called μ_N^*-measurable if f is measurable with respect to the tribe of all μ_N^*-measurable sets.

Theorem 16. – *Let $A \subset X$ be a set and $f: X \to \bar{R}$ a function. Consider the assertions:*

i) *A is integrable.*
ii) *A is μ_N^*-measurable and $\mu_N^*(A) < \infty$.*
iii) *A is measurable.*
iv) *A is μ_N^*-measurable.*
v) *f is measurable.*
vi) *f is μ_N^*-measurable.*

Then i) $\Leftrightarrow$ ii), iii) $\Leftrightarrow$ iv) *and* v) $\Leftrightarrow$ vi).

7. The essential upper integral. The spaces M^∞ and L^∞

For each $f \in \bar{R}_+^X$ we define

$$\bar{N}(f) = \sup\{N(gf) \mid g \in \mathscr{R}_+, g \leqslant 1\}.$$

It is easy to see that $\bar{N}: \bar{R}_+^X \to \bar{R}$ is an upper integral on X. We shall call $\bar{N}$ the *essential upper integral* associated with $(N, \mathscr{R})$.

Among the properties of $\bar{N}$ we would like to list the following:

(1) $\bar{N} \leqslant N$.

(2) For each $f \in \bar{R}_+^X$

$$\bar{N}(f) = \sup\{N(f\varphi_A) \mid A \in \mathscr{B}_0(X, N, \mathscr{R})\}.$$

(3) If $f \in \bar{R}_+^X$ vanishes on the complement of a countable union of sets belonging to $\mathscr{B}_0(X, N, \mathscr{R})$ then $\bar{N}(f) = N(f)$. In particular, $\bar{N}(f) = N(f)$ for $f \in \mathscr{L}^p(X, N, \mathscr{R})$ $(1 \leqslant p < \infty)$. Also $\bar{N}(f) = N(f)$ if $N(f) < \infty$.

(4) $\bar{N}$ is regular (see Definition 4, section 4).

(5) $(\bar{\bar{N}}) = \bar{N}$.

A set $A \subset X$ is called *locally N-negligible* if $A \cap B$ is N-negligible for each $B \in \mathscr{B}_0$. By property (2) above, $A \subset X$ is locally N-negligible if and only if $\bar{N}(\varphi_A) = 0$ (i.e., A is $\bar{N}$-negligible).

A function $f: X \to R$ is called *locally N-negligible* if $\{x \mid f(x) \neq 0\}$ is locally N-negligible. Hence $f: X \to \bar{R}$ is locally N-negligible if and only if $\bar{N}(|f|) = 0$, that is, $|f|$ is $\bar{N}$-negligible.

Theorem 17. – 17.1) *A function* $f: X \to \bar{R}$ *is* $(N, \mathscr{R})$*-measurable if and only if it is* $(\bar{N}, \mathscr{R})$*-measurable.* 17.2) *A set* $A \subset X$ *is* μ_N^**-measurable if and only if it is* $\mu_{\bar{N}}^*$*-measurable.*

Let $B^\infty(X)$ be the *algebra* of all *bounded* mappings of X into R. For each $f \in B^\infty(X)$ let

$$\|f\|_\infty = \sup_{t \in X} |f(t)|;$$

then $f \to \|f\|_\infty$ is a *norm* on $B^\infty(X)$ and $B^\infty(X)$ is a *Banach algebra* (when endowed with this norm).

For $f \in B^\infty(X)$ let A_f be the set of all $\alpha > 0$ such that the set

$$\{t \mid |f(t)| > \alpha\}$$

is locally N-negligible. If we define

$$N_\infty(f) = \inf A_f$$

then N_∞ is a *seminorm* on $B^\infty(X)$. Note that if $f \in B^\infty(X)$, then $N_\infty(f) = 0$ if and only if f *is locally* N*-negligible.*

We now denote by $M^\infty(X, N, \mathscr{R})$ the *algebra* of all *bounded* functions $f \in \mathscr{L}(X, N, \mathscr{R})$ and by $\mathscr{N}^\infty(X, N, \mathscr{R})$ the ideal of $M^\infty(X, N, \mathscr{R})$ consisting of all *bounded locally* N*-negligible functions.* By $L^\infty(X, N, \mathscr{R})$ we denote the *quotient algebra*

$$M^\infty(X, N, \mathscr{R}) / \mathscr{N}^\infty(X, N, \mathscr{R}).$$

When there is no ambiguity we shall write

M^∞ instead of $M^\infty(X, N, \mathscr{R})$;
L^∞ instead of $L^\infty(X, N, \mathscr{R})$;
$\mathscr{N}^\infty$ instead of $\mathscr{N}^\infty(N, X, \mathscr{R})$.

The restriction of N_∞ to M^∞ is clearly a seminorm on M^∞; we denote it by the same symbol. The corresponding *norm* on the quotient space L^∞ will be again denoted by N_∞; when endowed with this norm L^∞ is a *Banach algebra.*

We would like to note that *if* $f \in \mathscr{L}^1(X, N, \mathscr{R})$ *and* $g \in M^\infty(X, N, \mathscr{R})$ then $fg \in \mathscr{L}^1(X, N, \mathscr{R})$ *and*

$$|\int f g \, d\mu| \leqslant N_1(f) \, N_\infty(g).$$

Suppose now that:

e) $N = \bar{N}$.

In this case (with the notation of section 5) if $f: X \to R$ is bounded we have

$$f \in \mathscr{N}^\infty \Leftrightarrow f \equiv 0.$$

Hence the cannonical mapping of M^∞ onto L^∞ can be identified with the restriction of π to M^∞ and L^∞ can be identified with the image of M^∞ under π (we denote the canonical mapping of M^∞ onto L^∞ again by π).

We shall close this section with the following result (note that here we suppose $N=\bar{N}$):

Proposition 3. – *Let $\mathscr{C}\subset\mathscr{B}$. Then the following assertions are equivalent:*

i) $\sup\{\tilde{K}|K\in\mathscr{C}\}=\tilde{X}$.
ii) *For every $B\in\mathscr{B}_0$ there is a countable set $\mathscr{C}_B\subset\mathscr{C}$ such that*

$$B-\bigcup_{K\in\mathscr{C}_B} K$$

is negligible.

8. Localizable and strictly localizable spaces

We shall now give the following:

Definition 7. – *We say that $(X,N,\mathscr{R})$ is localizable if:*

i) $N=\bar{N}$;
ii) *$L(X,N,\mathscr{R})$ is completely reticulated.*

Note that whenever we suppose $(X,N,\mathscr{R})$ localizable we have $N=\bar{N}$.

Theorem 18[1]). – *Suppose $N=\bar{N}$. Then the following assertions are equivalent:*

i) *$(X,N,\mathscr{R})$ is localizable;*
ii) *L^∞ is completely reticulated;*
iii) *The Boolean algebra $\mathscr{B}/\mathscr{N}_0$ is complete;*
iv) *Let $\mathscr{C}\subset\mathscr{B}$ be such that $\sup\{\tilde{K}|K\in\mathscr{C}\}=\tilde{X}$. Let $(f_K)_{K\in\mathscr{C}}$ be a family of functions belonging to $\mathscr{L}$ with the property that for any $K'\in\mathscr{C}$, $K''\in\mathscr{C}$, $f_{K'}(x)=f_{K''}(x)$ almost everywhere on $K'\cap K''$. There is then $f\in\mathscr{L}$ such that, for every $K\in\mathscr{C}$, $f(x)=f_K(x)$ almost everywhere on K.*
v) *For every continuous linear functional u on L^1* (i.e., $u\in(L^1)'$) *there is $\tilde{g}_u\in L^\infty$ such that*

(1) $$u(\tilde{f})=\int f g_u\,d\mu \quad \text{for all} \quad f\in\mathscr{L}^1.$$

Remarks. – 1) With the notation of statement iv) above, if $f'\in\mathscr{L}$ is such that for every $K\in\mathscr{C}$, $f'(x)=f_K(x)$ almost everywhere on K,

[1]) Concerning Theorem 18 see also [126].

then $f' \equiv f$. 2) With the notation of statement v) above, for each $u \in (L^1)'$ the element $\tilde{g}_u \in L^\infty$ satisfying the relations (1) is uniquely determined by (1). The mapping $u \to \tilde{g}_u$ is an (isometric) isomorphism of the Banach space $(L^1)'$ onto the Banach space L^∞.

We shall need later a notion which is stronger than localizability. We define it as follows:

Definition 8. – *We say that* $(X, N, \mathcal{R})$ *is strictly localizable if:*

(i) $N = \bar{N}$;

(ii) *There exists a partition* $\mathcal{C}$ *of* X *consisting of non-negligible integrable sets such that* $\sup\{\tilde{K} \mid K \in \mathcal{C}\} = \tilde{X}$.

Proposition 3 in the previous section yields an equivalent formulation of strict localizability. Note also that if $(X, N, \mathcal{R})$ is *strictly localizable* then $(X, N, \mathcal{R})$ is *localizable* (use for instance Theorems 4, 15 and 18).

9. The case of abstract measures and of Radon measures

I) Let $(X, \mathcal{E}, \mu)$ be a *measure space*, i.e., X a set, $\mathcal{E}$ a tribe of subsets of X, μ a positive countably additive set function on $\mathcal{E}$.

Denote by $\mathcal{R}$ *the set of all* $\mathcal{E}$*-measurable simple functions* f *on* X *for which* $\mu(\{x \mid f(x) \neq 0\}) < \infty$.

Let $f \in \mathcal{R}$. Then f can be written in the form:

(*) $$f = \sum_{j \in J} c_j \varphi_{A_j}$$

where J is finite, $c_j \neq 0$, $A_j \in \mathcal{E}$ and $\mu(A_j) < +\infty$ for each $j \in J$. We define $\mu(f)$ be the equation

$$\mu(f) = \sum_{j \in J} c_j \mu(A_j).$$

It is easily seen that the definition of $\mu(f)$ does not depend on the particular representation of f of the form (*), that the mapping μ of $\mathcal{R}$ into R is linear, positive, and "continuous from above at 0 for sequences" (i.e., if (f_n) is a decreasing sequence of functions belonging to $\mathcal{R}_+$ which converges pointwise to zero, then $\inf_n \mu(f_n) = 0$).

We shall now define the upper integral N as follows: For each $f \in \bar{R}_+^X$ let $\mathcal{R}_f$ be the set consisting of all countable families (g_n) with the following two properties:

$$g_n \in \mathcal{R}_+ \quad \text{for each } n \text{ and } \quad \sum_n g_n \geq f.$$

We define $N(f)$ by the equations:

$$N(f) = \begin{cases} \inf\left\{\sum_n \mu(g_n) \,\middle|\, (g_n) \in \mathscr{R}_f\right\} & \text{if } \mathscr{R}_f \neq \varnothing, \\ +\infty \quad \text{if } \mathscr{R}_f = \varnothing. \end{cases}$$

We note that for $g \in \mathscr{R}_+$, $N(g) = \mu(g)$. It can be verified that $N: \bar{R}_+^X \to \bar{R}_+$ is an *upper integral on* X, that $\mathscr{R}$ *and* N *satisfy the axioms* (L_1), (L_2), (L_3) in section 2 and finally that *the upper integral* N *is regular.*

Moreover if X is a countable union of sets $A \in \mathscr{E}$ satisfying $\mu(A) < +\infty$, then $(X, N, \mathscr{R})$ is *strictly localizable.*

II) Let X be a *locally compact space.* Let $\mathscr{K}(X)$[1]) be the *set of all continuous mappings of* X *into* R *having compact support.* Let μ be a *positive Radon measure* on X, that is $\mu: \mathscr{K}(X) \to R$ is a positive linear mapping.

It is well-known (see chapters 1–4, [15]) that μ is "continuous from above at 0 for directed families" (i.e., if (f_α) is a family of functions belonging to $\mathscr{K}_+(X)$, directed for the relation $\geqslant$ and converging pointwise to zero, then $\inf_\alpha \mu(f_\alpha) = 0$).

Let $\mathfrak{I}_+(X)$ be the set of all mappings of X into $\bar{R}_+$ which are lower semi-continuous. We now recall Bourbaki's construction of the upper integral associated with the positive Radon measure μ. The construction is done in two steps. First μ is extended to $\mathfrak{I}_+(X)$. For $g \in \mathfrak{I}_+(X)$ we define

$$N(g) = \sup\{\mu(f) \mid f \in \mathscr{K}_+(X),\ f \leqslant g\}.$$

It is easily seen that $N(g) = \mu(g)$ if $g \in \mathscr{K}_+(X)$, that N is positively homogeneous and additive on $\mathfrak{I}_+(X)$. Moreover N has the following important and useful "continuity property": If (g_α) is a family of functions belonging to $\mathfrak{I}_+(X)$, directed for the relation $\leqslant$ and if g_∞ is the upper envelope of (g_α), then (clearly $g_\infty \in \mathfrak{I}_+(X)$) $N(g_\infty) = \sup_\alpha N(g_\alpha)$.

The second and final step in the construction of the upper integral consists of extending the definition of N from $\mathfrak{I}_+(X)$ to $\bar{R}_+^X$. For $f \in \bar{R}_+^X$ we define

$$N(f) = \inf\{N(g) \mid g \in \mathfrak{I}_+(X),\ g \geqslant f\}.$$

It is easy to verify that $N: \bar{R}_+^X \to \bar{R}_+$ is an *upper integral* on X, that $\mathscr{R} = \mathscr{K}(X)$ *and* N *satisfy the axioms* (L_1), (L_2), (L_3) in section 2 and finally that the *upper integral* N *is regular.*

We would also like to point out that Bourbaki's essential upper integral as defined in chapter 5, § 1, [15] coincides with our essential upper integral as defined in section 7 above.

---

[1]) Here $\mathscr{K}(X)$ plays the role of $\mathscr{R}$ in section 2.

If N is the upper integral associated with μ, we shall often write

$$\int_X^* f\,d\mu = N(f)$$

and

$$\int_X^\bullet f\,d\mu = \bar{N}(f)$$

for $f \in \bar{R}_+^X$.

We end this section with the following remark: It follows from Proposition 14, chapter 4, § 5, [15] that $(X, \bar{N}, \mathscr{K}(X))$ *is always stricly localizable.*

CHAPTER II

Admissible subalgebras and projections onto them

In this chapter we give several measure-theoretical results. In particular we define the notion of admissible subalgebra of M^∞ and we study the projections onto such subalgebras.

Throughout this chapter we assume that $(X, N, \mathscr{R})$ *is localizable.*

1. Admissible subalgebras

We shall start with the following:

Proposition 1. – *Let* $\mathscr{A}$ *be a subalgebra of* M^∞ *containing* 1 *and closed under pointwise convergence of bounded sequences. Then* $u \circ g \in \mathscr{A}$ *for every bounded Borel measurable function* $u: R \to R$ *and every* $g \in \mathscr{A}$. *In particular:*

1) $|g| \in \mathscr{A}$ *for each* $g \in \mathscr{A}$;
2) $\mathscr{A}$ *is a Riesz space (for the usual order relation);*
3) $\mathscr{A}$ *is closed under* lim sup *and* lim inf *of bounded sequences.*

Proof: We only need to show that if $u: R \to R$ is bounded Borel measurable and $g \in \mathscr{A}$, then $u \circ g \in \mathscr{A}$ (statements 1), 2), 3) then follow immediately). Let $g \in \mathscr{A}$, $g \neq 0$, let $a = \|g\|_\infty$ and let

$$\mathscr{H} = \{h: [-a, a] \to R \mid h \quad \text{bounded and} \quad h \circ g \in \mathscr{A}\}.$$

Clearly $\mathscr{H}$ is an algebra and $\mathscr{H}$ is closed under pointwise convergence of bounded sequences. Since $\mathscr{H}$ contains the polynomial functions on $[-a, a]$, $\mathscr{H}$ contains also $C([-a, a])$ (use the Stone-Weierstrass theorem) and therefore $\mathscr{H}$ contains the bounded Borel measurable functions on $[-a, a]$. Let now $u: R \to R$ be bounded Borel measurable. Then $u|[-a, a] \in \mathscr{H}$ and hence $u \circ g = (u|[-a, a]) \circ g \in \mathscr{A}$.

Proposition 2. – *Let* $\mathscr{A}$ *be a subalgebra of* M^∞ *containing* 1 *and let* $\mathscr{F} = \{A \mid \varphi_A \in \mathscr{A}\}$. *The following assertions are equivalent:*

i) $\mathscr{A}$ *is closed under pointwise convergence of bounded sequences.*

ii) $\mathscr{F}$ *is a tribe of subsets of* X *and a bounded function* $f: X \to R$ *belongs to* $\mathscr{A}$ *if and only if* f *is* $\mathscr{F}$*-measurable.*

Proof: i)$\Rightarrow$ii). Let us show that $\mathscr{F}$ is a tribe. For $A \in \mathscr{F}$ we have:

$$\varphi_A \in \mathscr{A} \Rightarrow \varphi_{CA} = 1 - \varphi_A \in \mathscr{A} \Rightarrow CA \in \mathscr{F}.$$

If (A_n) is a sequence of sets belonging to $\mathscr{F}$, then by 3), Proposition 1,

$$\varphi_{\bigcup_n A_n} \in \mathscr{A} \Rightarrow \bigcup_n A_n \in \mathscr{F}.$$

Thus $\mathscr{F}$ is a tribe. Since every simple function based on sets in $\mathscr{F}$ belongs to $\mathscr{A}$, it follows that every bounded $\mathscr{F}$-measurable function belongs to $\mathscr{A}$. Conversely, let $f \in \mathscr{A}$, $a \in R$ and $E = \{x | f(x) > a\}$. Then φ_E is the pointwise limit of the sequence $(\inf(n(f-a)^+, 1))$ and for each n, $\inf(n(f-a)^+, 1) \in \mathscr{A}$. Consequently $E \in \mathscr{F}$.

Since ii)$\Rightarrow$i) obviously, this completes the proof.

Remark. – Let $\mathscr{A}$ be a subset of M^∞ and let $\mathscr{F} = \{A | \varphi_A \in \mathscr{A}\}$. If $\mathscr{N}^\infty \subset \mathscr{A}$, then $\mathscr{N}_0 \subset \mathscr{F}$.

Below if $A \subset L^2$ then $\bar{A}$ denotes the closure of A in L^2; we also recall that we denote by π the canonical mapping of M^∞ onto L^∞.

Proposition 3. – *Let* $\mathscr{A}$ *be a subalgebra of* M^∞ *and let*

$$\mathscr{F}_0 = \{F | F \in \mathscr{B}_0, \varphi_F \in \mathscr{A}\}.$$

We suppose that $\mathscr{A}$ *contains* 1 *and* $\mathscr{N}^\infty$. *Consider the following assertions:*

α) $\mathscr{A}$ *is closed under pointwise convergence of bounded sequences.*

β) *If* $f: X \to R$ *is a bounded function such that* $f\varphi_F \in \mathscr{A}$ *for each* $F \in \mathscr{F}_0$, *then* $f \in \mathscr{A}$.

α') *If* $H \subset \pi(\mathscr{A})_+$ *is a bounded set and if* $\tilde{h}_\infty$ *is the supremum*[1]) *of* H, *then* $\tilde{h}_\infty \in \pi(\mathscr{A})$. *In particular* $\pi(\mathscr{A})$ *is a completely reticulated Riesz space.*

β') $\sup\{\tilde{F} | F \in \mathscr{F}_0\} = \tilde{X}$.

γ) $\overline{\pi(\mathscr{A}) \cap L^2} \cap L^\infty = \pi(\mathscr{A}) \cap L^2$.

Then $\{\alpha), \beta)\} \Leftrightarrow \{\alpha'), \beta')\} \Leftrightarrow \{\beta), \gamma)\}$.

Proof: It suffices to prove the chain of implications

$$\{\alpha), \beta)\} \Rightarrow \{\beta), \gamma)\} \Rightarrow \{\alpha'), \beta')\} \Rightarrow \{\alpha), \beta)\}.$$

$\{\alpha), \beta)\} \Rightarrow \{\beta), \gamma)\}$. It is clearly enough to show that $\{\alpha), \beta)\} \Rightarrow \gamma)$. Let $\tilde{h} \in \overline{\pi(\mathscr{A}) \cap L^2} \cap L^\infty$. There is then a sequence $(\tilde{h}_n)$ of elements of

[1]) Unless explicitly stated otherwise, the "sup" and "inf" are always computed in $\mathscr{B}/\mathscr{N}_0$ and L, respectively.

$\pi(\mathscr{A})\cap L^2$ such that $\lim_n N_2(\tilde{h}-\tilde{h}_n)=0$. Let $a=N_\infty(\tilde{h})$. We may assume that $|h|\leqslant a$, that $\lim_n h_n(x)=h(x)$ for all $x\in X$, and that $|h_n|\leqslant a$ for every n (use the fact that $\mathscr{A}$ is a Riesz space). We deduce that $h\in\mathscr{A}$; since $h\in\mathscr{L}^2$ we conclude that $\tilde{h}\in\pi(\mathscr{A})\cap L^2$. Hence

$$\overline{\pi(\mathscr{A})\cap L^2}\cap L^\infty\subset\pi(\mathscr{A})\cap L^2.$$

Since the converse inclusion is obvious, the implication $\{\alpha), \beta)\}\Rightarrow\{\beta), \gamma)\}$ is proved.

$\{\beta), \gamma)\}\Rightarrow\{\alpha'), \beta')\}$. We note first that $\beta)\Rightarrow\beta')$ In fact let $\tilde{Y}=\sup\{\tilde{F}|F\in\mathscr{F}_0\}$ and suppose $\tilde{Y}\neq\tilde{X}$. Since $\complement Y$ is measurable and non-negligible, there exists B integrable and non-negligible such that $B\subset\complement Y$ (use the fact that $\bar{N}=N$). Clearly $B\notin\mathscr{F}_0$ and thus $\varphi_B\notin\mathscr{A}$; however $\varphi_B\varphi_F\equiv 0$ and hence $\varphi_B\varphi_F\in\mathscr{A}$ for each $F\in\mathscr{F}_0$. Thus we reached a contradiction and hence $\beta)\Rightarrow\beta')$ is proved. It remains to show that $\{\beta), \gamma)\}\Rightarrow\alpha')$. We note first that if $H\subset\pi(\mathscr{A})$ is a finite set, then $\sup H$ belongs to $\pi(\mathscr{A})$. For this it is enough to show that if $\tilde{f}\in\pi(\mathscr{A})$ then $|\tilde{f}|\in\pi(\mathscr{A})$. Let then $f\in\mathscr{A}$ and let $A>0$ such that $f(X)\subset[-A,A]$. Let (p_n) be a sequence of polynomials converging uniformly on $[-A,A]$ to the function $x\to|x|$; we may suppose $p_n(0)=0$ for all n. Let now $F\in\mathscr{F}_0$. Then $f\varphi_F\in\mathscr{A}\cap\mathscr{L}^2$, whence $p_n(f\varphi_F)=\varphi_F p_n(f)$ belongs to $\mathscr{A}\cap\mathscr{L}^2$ for all n. We deduce that

$$|\tilde{f}|\,\tilde{\varphi}_F\in\overline{\pi(\mathscr{A})\cap L^2}\cap L^\infty=\pi(\mathscr{A})\cap L^2\subset\pi(\mathscr{A}).$$

Hence $|f|\,\varphi_F\in\mathscr{A}$. Since $F\in\mathscr{F}_0$ was arbitrary we conclude that $|f|\in\mathscr{A}$. Let now $H\subset\pi(\mathscr{A})_+$ be a bounded set directed for the relation $\leqslant$ and let $\tilde{h}_\infty=\sup H$. For each $F\in\mathscr{F}_0$ let

$$H_F=\{\tilde{h}\,\tilde{\varphi}_F|\tilde{h}\in H\}.$$

Clearly $H_F\subset\pi(\mathscr{A})\cap L^2_+$ is a bounded set both in L^2 and in L^∞ and $\sup H_F=\tilde{h}_\infty\,\tilde{\varphi}_F$. By Theorem 4, section 2, chapter 1, we have

$$\tilde{h}_\infty\,\tilde{\varphi}_F\in\overline{\pi(\mathscr{A})\cap L^2_+}\subset\overline{\pi(\mathscr{A})\cap L^2}.$$

Since $\tilde{h}_\infty\,\tilde{\varphi}_F$ is clearly also in L^∞ we conclude

$$\tilde{h}_\infty\,\tilde{\varphi}_F\in\overline{\pi(\mathscr{A})\cap L^2}\cap L^\infty=\pi(\mathscr{A})\cap L^2.$$

Hence $h_\infty\varphi_F\in\mathscr{A}$. Since $F\in\mathscr{F}_0$ was arbitrary, condition $\beta)$ implies that $h_\infty\in\mathscr{A}$, i.e. $\tilde{h}_\infty\in\pi(\mathscr{A})$. The statement about $\pi(\mathscr{A})$ being a completely reticulated Riesz space now follows immediately.

$\{\alpha'), \beta')\}\Rightarrow\{\alpha), \beta)\}$. We note first that $\alpha')\Rightarrow\alpha)$. In fact it is enough to remark that for every bounded sequence (f_n) of functions in $\mathscr{A}$, the pointwise supremum and the pointwise infimum of (f_n) "belong" to $\sup_n \tilde{f}_n$ and $\inf_n \tilde{f}_n$, and hence are in $\mathscr{A}$. It remains to show that $\{\alpha'), \beta')\} \Rightarrow \beta)$. Let $f: X \to R$ be a bounded function such that $f\varphi_F \in \mathscr{A}$ for each $F \in \mathscr{F}_0$. We may assume without loss of generality that $f \geqslant 0$. From β') and Theorem 15, section 5, chapter 1, we deduce that f is measurable. By Theorem 14, section 5, chapter 1, we have $\tilde{f} = \sup\{\tilde{f}\tilde{\varphi}_F | F \in \mathscr{F}_0\}$. Condition α') then implies that $f \in \mathscr{A}$.

This completes the proof of Proposition 3.

Remarks. 1) Roughly speaking condition β) of Proposition 3 expresses the fact that "there are enough integrable sets in $\mathscr{F} = \{A | \varphi_A \in \mathscr{A}\}$ to characterize the notion of measurable function belonging to $\mathscr{A}$".

2) If X is integrable then conditions β) and β') are automatically satisfied.

We shall now give one of our main definitions:

Definition 1. – *A subalgebra $\mathscr{A}$ of M^∞ is called admissible if:*

i) $1 \in \mathscr{A}, \mathscr{N}^\infty \subset \mathscr{A}$;
ii) *$\mathscr{A}$ satisfies $\{\alpha), \beta)\}$ of Proposition* 3.

We recall that we just proved that $\{\alpha), \beta)\}$, $\{\alpha'), \beta')\}$ and $\{\beta), \gamma)\}$ are equivalent.

Let $\mathscr{A} \subset M^\infty$ be an admissible algebra and let $\mathscr{F} = \{A | \varphi_A \in \mathscr{A}\}$ and $\mathscr{F}_0 = \{F | F \in \mathscr{B}_0, \varphi_F \in \mathscr{A}\}$. We have:

Remarks. – 1) If $\tilde{u} \in \overline{\pi(\mathscr{A}) \cap L^2}$, then u is $\mathscr{F}$-measurable.

2) For each $f: X \to \bar{R}_+$ we have

$$N(f) = \sup\{N(f\varphi_F) | F \in \mathscr{F}_0\}$$

(use condition β') of Proposition 3 above, the fact that

$$N(f) = \sup\{N(f\varphi_B) | B \in \mathscr{B}_0\}$$

and Proposition 3, section 7, chapter 1).

3) Let $h \in \mathscr{L}^p$ $(1 \leqslant p < \infty)$ be $\mathscr{F}$-measurable. If $\int h \cdot \varphi_F \, d\mu \geqslant 0$ for every $F \in \mathscr{F}_0$, then $h(x) \geqslant 0$ almost everywhere (use β') of Proposition 3). In particular if $\int h\varphi_F \, d\mu = 0$ for all $F \in \mathscr{F}_0$, then $h \equiv 0$.

4) Let $h \in \mathscr{L}^p$ $(1 \leqslant p < \infty)$ be $\mathscr{F}$-measurable and suppose that

$$\sup\left\{\left|\int hg\,d\mu\right| \,|\, g \in \mathscr{A} \text{ simple integrable with } N_\infty(g) \leqslant 1\right\}$$

is finite. Then $h \in \mathscr{L}^1$ and $N_1(h)$ is equal to the above supremum (use Remark 2) above).

5) Let $\mathscr{C} \subset \mathscr{F}_0$ be a set with the property that $\sup\{\tilde{K} | K \in \mathscr{C}\} = \tilde{X}$. We have:

5.1) If $f: X \to R$ is a bounded function such that $f\varphi_K \in \mathscr{A}$ for each $K \in \mathscr{C}$, then $f \in \mathscr{A}$ (use Theorems 14, 15, section 5, chapter 1, and Proposition 3 above).

5.2) If $f: X \to R$ is a bounded function such that $f\varphi_K \equiv 0$ for each $K \in \mathscr{C}$, then $f \equiv 0$ (use Theorem 15, section 5, chapter 1).

2. Multiplicative linear mappings

In this section we give a theorem of R. R. Phelps concerning multiplicative linear mappings (see [63], [65], [113]). Since we shall make use of this result again in later chapters, we state it here in the form best suited for our purposes.

We recall that an algebra A (over R) endowed with an order relation $\leqslant$ is called an *ordered algebra* if:

i) A is an ordered vector space with respect to $\leqslant$; ii) the relations $0 \leqslant x$, $0 \leqslant y$ imply $0 \leqslant xy$.

Let now A and B be two ordered algebras (over R) and having *strictly positive* unit elements (we denote the order relations in A and B by $\leqslant$ and the unit elements in A and B by 1). We suppose in addition that every element of A is *bounded*: for each $x \in A$ there is a positive real number λ such that $-\lambda 1 \leqslant x \leqslant \lambda 1$ (note that this implies $A = A_+ - A_+$). Let $\mathscr{L}^*(A,B)$ be the vector space of all linear mappings of A into B and define

$$\mathscr{L}_1^*(A,B) = \{T \in \mathscr{L}^*(A,B) \mid T \geqslant 0^{1)} \text{ and } T1 = 1\}.$$

Note that $\mathscr{L}_1^*(A,B)$ is a *convex* set. With these notations we have:

Theorem 1 (R. R. Phelps).[2]) – *If $T \in \mathscr{L}_1^*(A,B)$ is an extremal point of $\mathscr{L}_1^*(A,B)$, then T is multiplicative.*

Proof: Let $u \in A$, $0 \leqslant u \leqslant 1$ and define $T^{(u)}: A \to B$ by

$$T^{(u)}x = T(xu) - Tx \cdot Tu, \qquad x \in A.$$

Clearly $T^{(u)} \in \mathscr{L}^*(A,B)$. It is easily seen that $T + T^{(u)}$ and $T - T^{(u)}$ belong to $\mathscr{L}_1^*(A,B)$. Let us verify for instance that $T + T^{(u)} \in \mathscr{L}_1^*(A,B)$. For $x \in A$, $x \geqslant 0$, we have

$$(T + T^{(u)})x = Tx + T(xu) - Tx \cdot Tu = (Tx)(1 - Tu) + T(xu) \geqslant 0$$

---

[1]) $T \geqslant 0$ means that $x \in A$ and $x \geqslant 0$ imply $Tx \geqslant 0$.

[2]) This theorem was previously proved (and used by the authors for $A = C(Z_1)$, $B = C(Z_2)$, Z_1 and Z_2 compact (see [63], [65]).

(since $0 \leqslant Tu \leqslant T1 = 1$). We also have

$$(T+T^{(u)})1 = T1 + T(1u) - T1 \cdot Tu = 1 + Tu - Tu = 1.$$

Thus $T+T^{(u)} \in \mathscr{L}_1^*(A,B)$ as asserted (the proof for $T-T^{(u)}$ is entirely similar).

Since $T = \frac{1}{2}(T+T^{(u)}) + \frac{1}{2}(T-T^{(u)})$ and T is extremal in $\mathscr{L}_1^*(A,B)$, we deduce $T+T^{(u)} = T-T^{(u)}$, whence $T^{(u)} = 0$. This means that

$$T(xu) = Tx \cdot Tu \quad \text{for all} \quad x \in A.$$

By linearity we deduce that the previous identities remain valid for any $u \in A$, $u \geqslant 0$ and finally for arbitrary $u \in A$. This shows that T is multiplicative and hence completes the proof of the theorem.

Throughout the remainder of this section $\mathscr{A}$ will be a *fixed admissible subalgebra* of M^∞. Note that $\pi(\mathscr{A})$ is a closed subalgebra of L^∞. Hence (with respect to the N_∞ norm) $\pi(\mathscr{A})$ is a commutative Banach algebra (over R) with unit element. Let $(\pi(\mathscr{A}))'$ be the Banach space dual of $\pi(\mathscr{A})$ and define:

$$P = \{x' \in (\pi(\mathscr{A}))' \mid x'(\tilde{f}) \geqslant 0 \quad \text{if} \quad \tilde{f} \geqslant \tilde{0} \quad \text{and} \quad x'(\tilde{1}) = 1\}.$$

We recall the following well-known result:[1])

Proposition 4. – *An element $x' \in P$ is an extremal point of P if and only if x' is a character of $\pi(\mathscr{A})$.*

For completeness we show how Proposition 4 can be obtained as a corollary of Phelps' theorem. To show that every extremal point of P is a character, we apply Theorem 1 with $A = \pi(\mathscr{A})$ and $B = R$. The hypotheses of Theorem 1 are satisfied and P coincides with $\mathscr{L}_1^*(A,B)$.

Conversely, assume that $x' \in P$ is a character and that $x' = \lambda y' + (1-\lambda) z'$ with $y' \in P$, $z' \in P$ and $0 < \lambda < 1$. Note that $x'(\tilde{\varphi}_E)$ is 0 or 1 for each E in $\mathscr{F} = \{A \mid \varphi_A \in \mathscr{A}\}$, since x' is multiplicative, and that $0 \leqslant y'(\tilde{\varphi}_E) \leqslant 1$, $0 \leqslant z'(\tilde{\varphi}_E) \leqslant 1$ for each $E \in \mathscr{F}$. We deduce that

$$x'(\tilde{\varphi}_E) = 1 \Rightarrow y'(\tilde{\varphi}_E) = z'(\tilde{\varphi}_E) = 1$$

and

$$x'(\tilde{\varphi}_E) = 0 \Rightarrow y'(\tilde{\varphi}_E) = z'(\tilde{\varphi}_E) = 0,$$

that is $x'(\tilde{\varphi}_E) = y'(\tilde{\varphi}_E) = z'(\tilde{\varphi}_E)$ for all $E \in \mathscr{F}$. It follows that x', y', z' coincide on "every simple function based on sets belonging to $\mathscr{F}$". By continuity we deduce that $x' = y' = z'$. This shows that x' is extremal in P and completes the proof.

From Theorem 1 and Proposition 4 we deduce the following:

Corollary – *Let B be an ordered algebra with strictly positive unit element. Suppose that the positive characters of B separate the elements*

---

[1]) If Q is an algebra (over R), then a character of Q is any multiplicative linear mapping $\chi: Q \to R$ such that $\chi \neq 0$.

of B. Then an element $T \in \mathscr{L}_1^*(\pi(\mathscr{A}),B)$ is extremal in $\mathscr{L}_1^*(\pi(\mathscr{A}),B)$ if and only if it is multiplicative.

Proof: By Theorem 1, T is multiplicative if T is extremal.

Conversely, suppose that T is multiplicative and that $T=\lambda T_1 +(1-\lambda)T_2$ with $T_1 \in \mathscr{L}_1^*(\pi(\mathscr{A}),B)$, $T_2 \in \mathscr{L}_1^*(\pi(\mathscr{A}),B)$ and $0<\lambda<1$. Let now χ be a positive character of B. Then $\chi \circ T$, $\chi \circ T_1$ and $\chi \circ T_2$ belong to P,

$$\chi \circ T = \lambda \chi \circ T_1 + (1-\lambda)\chi \circ T_2$$

and $\chi \circ T$ is a character of $\pi(\mathscr{A})$. By Proposition 4, $\chi \circ T$ is extremal in P and hence $\chi \circ T = \chi \circ T_1 = \chi \circ T_2$. Since χ was arbitrary we deduce that $T=T_1=T_2$ and hence T is extremal.

Remark. – For further results concerning multiplicative linear mappings and extremal points see [13], [37].

3. Extensions of linear mappings

In this section we assume that $\mathscr{C}$ is a *fixed collection of non-negligible integrable sets such that* $\sup\{\tilde{K} | K \in \mathscr{C}\} = \tilde{X}$ and that $E \subset L^\infty$ *is a vector space such that* $\tilde{\varphi}_K L^\infty \subset E$ *for every* $K \in \mathscr{C}$.

Let $T: E \to E$ be a linear mapping and let $\tilde{u} \in L^\infty$. We say that T *commutes with* $\tilde{u}$ if $\tilde{u} \cdot E \subset E$ and $T(\tilde{u}\tilde{f}) = \tilde{u}T\tilde{f}$ for all $\tilde{f} \in E$; clearly if T commutes with $\tilde{u}_1 \in L^\infty$ and with $\tilde{u}_2 \in L^\infty$ then T commutes with $\tilde{u}_1 \tilde{u}_2$.

Let F be a vector space and $P: F \to F$. We say that P is a *projection* if P is linear and $P^2 = P$. Note that $Px = x$ if and only if $x \in P(F)$.

Proposition 5. – *Let $T: E \to E$ be a continuous*[1]) *linear mapping commuting with $\tilde{\varphi}_K$ for every $K \in \mathscr{C}$. Then there exists a unique linear mapping $T': L^\infty \to L^\infty$ commuting with $\tilde{\varphi}_K$ for every $K \in \mathscr{C}$ and such that*

$$T' | E = T. \tag{1}$$

Moreover: 1) *T' is a projection if and only if T is a projection.*

2) *T' is positive if and only if T is positive.*

3) *If T commutes with $\tilde{u} \in L^\infty$ then T' commutes with $\tilde{u}$.*

4) *T' is continuous and has norm equal to the norm of T.*

Proof: We first prove the existence of a linear mapping $T': L^\infty \to L^\infty$ commuting with $\tilde{\varphi}_K$ for every $K \in \mathscr{C}$ and satisfying (1).

---

[1]) We consider on E the structure of normed space induced by that of L^∞.

Let $\tilde{f} \in L^\infty$ and let K and H be sets belonging to $\mathscr{C}$. Note that[1])

$$(2) \qquad T(\varphi_K f)(x) = T(\varphi_H f)(x)$$

almost everywhere on $K \cap H$. In fact we have

$$\tilde{\varphi}_{K \cap H} T(\tilde{\varphi}_K \tilde{f}) = T(\tilde{\varphi}_{K \cap H} \tilde{\varphi}_K \tilde{f}) = T(\tilde{\varphi}_{K \cap H} \tilde{f}) = T(\tilde{\varphi}_{K \cap H} \tilde{\varphi}_H \tilde{f}) = \tilde{\varphi}_{K \cap H} T(\tilde{\varphi}_H \tilde{f})$$

and hence (2) is proved. By Theorem 18, section 8, chapter 1 there exists a $g_f \in M^\infty$ (the class of which is uniquely determined) satisfying

$$(3) \qquad N_\infty(g_f) \leqslant \|T\| N_\infty(f) \quad \text{and} \quad g_f(x) = T(\varphi_K f)(x)$$

almost everywhere on K, for every $K \in \mathscr{C}$. We now define

$$T' \tilde{f} = \tilde{g}_f$$

for every $\tilde{f} \in L^\infty$. Clearly T' is well defined as a mapping of L^∞ into L^∞ and T' is linear. If $\tilde{f} \in E$ then the function $Tf (= g_f)$ satisfies (3), whence $T'|E = T$.

Let $K \in \mathscr{C}$. For every $\tilde{f} \in L^\infty$ we have

$$\tilde{\varphi}_K T' \tilde{f} = \tilde{\varphi}_K T(\tilde{\varphi}_K \tilde{f}) = T(\tilde{\varphi}_K \tilde{f}) = T'(\tilde{\varphi}_K \tilde{f});$$

therefore T' commutes with $\tilde{\varphi}_K$.

Let now T_1 and T_2 be two linear mappings of L^∞ into L^∞ commuting with $\tilde{\varphi}_K$ for every $K \in \mathscr{C}$ and satisfying

$$T_1|E = T_2|E.$$

Then $T_1 = T_2$. In fact let $\tilde{f} \in L^\infty$; for every $K \in \mathscr{C}$ we have

$$\tilde{\varphi}_K T_1 \tilde{f} = T_1(\tilde{\varphi}_K \tilde{f}) = T_2(\tilde{\varphi}_K \tilde{f}) = \tilde{\varphi}_K T_2 \tilde{f}$$

and theorem $T_1 \tilde{f} = T_2 \tilde{f}$.

It remains to prove the assertions 1), 2), 3), and 4).

1) If T' is a projection it follows immediately from (1) that T is a projection. Conversely suppose that T is a projection and let $\tilde{f} \in L^\infty$. For every $K \in \mathscr{C}$ we have

$$\tilde{\varphi}_K T'^2(\tilde{f}) = \tilde{\varphi}_K T'(T' \tilde{f}) = T'(T'(\tilde{\varphi}_K \tilde{f})) = T(T(\tilde{\varphi}_K \tilde{f})) = T(\tilde{\varphi}_K \tilde{f}) = \tilde{\varphi}_K T' \tilde{f}.$$

Therefore $T'^2 = T'$.

The proof of 2) is analogous and is left to the reader.

3) Suppose that T commutes with $\tilde{u} \in L^\infty$. Let $\tilde{f} \in L^\infty$. For every $K \in \mathscr{C}$ we have

$$\tilde{\varphi}_K T'(\tilde{u} \tilde{f}) = T(\tilde{\varphi}_K \tilde{u} \tilde{f}) = \tilde{u} T(\tilde{\varphi}_K \tilde{f}) = \tilde{u} \tilde{\varphi}_K T' \tilde{f}.$$

---

[1]) If $\Delta \subset L$ and $S: \Delta \to L$ then for $\tilde{f} \in \Delta$ we denote by Sf a representative of the class $S\tilde{f}$. If $S\tilde{f} \in L^\infty$ we choose Sf in M^∞.

We deduce that $\tilde{u}$ commutes with T'.

4) The assertion follows immediately from (3).

This completes the proof of Proposition 5.

Note concerning the terminology. – The mapping T' introduced in Proposition 5 will be called the *canonical extension* of T to L^∞. When there is no ambiguity we denote the canonical extension of T to L^∞ by the same letter T.

A *Dunford-Schwartz operator* is a linear mapping $T: L^1 \cap L^\infty \to L^1 \cap L^\infty$ such that $\|T\|_1 \leqslant 1$ and $\|T\|_\infty \leqslant 1$[1]. By the Riesz convexity theorem, $\|T\|_p \leqslant 1$ for each $1 \leqslant p < \infty$ and hence T can be extended by continuity to L^p (we denote the extension by the same letter). We denote by $\mathscr{D}$ the set of all Dunford-Schwartz operators and by $\mathscr{D}_+$ the set of positive Dunford-Schwartz operators.

Proposition 6. – *Let $P \in \mathscr{D}_+$ be an operator commuting with $\tilde{\varphi}_K$ for every $K \in \mathscr{C}$ (and consider its canonical extension to L^∞). If (f_n) is a bounded sequence in M^∞ converging pointwise to a function f, then (Pf_n) converges almost everywhere to Pf.*

For completeness we sketch a proof of this result.

Assume first that the sequence (f_n) is increasing. For each $K \in \mathscr{C}$, $(\varphi_K f_n)$ is an increasing sequence in $\mathscr{L}^1$ converging pointwise to $\varphi_K f \in \mathscr{L}^1$. By the monotone convergence theorem (Theorem 5, section 2, chapter 1), since P is continuous in L^1 and since the sequence $(P(\varphi_K f_n))$ is increasing, we deduce that $(P(\varphi_K f_n))$ converges to $P(\varphi_K f)$ almost everywhere, or equivalently that (Pf_n) converges to Pf almost everywhere on K. Since $K \in \mathscr{C}$ was arbitrary it follows that (Pf_n) converges almost everywhere to Pf.

The case of an arbitrary bounded sequence (f_n) in M^∞ converging pointwise to a function f can be readily reduced to the one discussed above.

4. Projections onto admissible subalgebras

In this section we define the notion of projection onto an admissible subalgebra (conditional expectation) and we show how to construct one using Hilbert space methods.

Throughout this section $\mathscr{A}$ is a *fixed admissible algebra.* As usual we write $\mathscr{F} = \{A \mid \varphi_A \in \mathscr{A}\}$ and $\mathscr{F}_0 = \{F \mid F \in \mathscr{B}_0,\ \varphi_F \in \mathscr{A}\}$. We recall that $\sup\{\tilde{F} \mid F \in \mathscr{F}_0\} = \tilde{X}$.

If $A \subset L^2$ then $\bar{A}$ denotes the closure of A in L^2.

---

[1]) If T is a Dunford-Schwartz operator and $1 \leqslant p \leqslant \infty$ we define

$$\|T\|_p = \sup\{N_p(T\tilde{f}) \mid \tilde{f} \in L^1 \cap L^\infty, \quad N_p(\tilde{f}) \leqslant 1\}.$$

Proposition 7. – *There exists one and only one linear mapping $P_{\mathscr{A}}$ of $L^2 \cap L^\infty$ into $\overline{L^2 \cap \pi(\mathscr{A})}$ satisfying the condition*

$$(*) \int P_{\mathscr{A}} f \cdot g \, d\mu = \int f \cdot g \, d\mu \quad \textit{for all} \quad \tilde{f} \in L^2 \cap L^\infty, \tilde{g} \in L^2 \cap \pi(\mathscr{A}).$$

The linear mapping $P_{\mathscr{A}}$ has the following properties:

1) *$P_{\mathscr{A}}$ is positive.*
2) *$N_\infty(P_{\mathscr{A}} \tilde{f}) \leqslant N_\infty(\tilde{f})$ for each $\tilde{f} \in L^2 \cap L^\infty$.*
3) *$P_{\mathscr{A}}$ maps $L^2 \cap L^\infty$ onto $L^2 \cap \pi(\mathscr{A})$ and $P_{\mathscr{A}}^2 = P_{\mathscr{A}}$ (in particular $P_{\mathscr{A}} \tilde{f} = \tilde{f}$ for every $\tilde{f} \in L^2 \cap \pi(\mathscr{A})$).*
4) *$P_{\mathscr{A}}(\tilde{g}\tilde{f}) = \tilde{g} P_{\mathscr{A}} \tilde{f}$ for all $\tilde{f} \in L^2 \cap L^\infty$ and $\tilde{g} \in L^2 \cap \pi(\mathscr{A})$.*

Proof: Let T be the orthogonal projection of L^2 onto $\overline{L^2 \cap \pi(\mathscr{A})}$ and let $P_{\mathscr{A}} = T | L^2 \cap L^\infty$. Then $P_{\mathscr{A}}: L^2 \cap L^\infty \to \overline{L^2 \cap \pi(\mathscr{A})}$ is a linear mapping. Since T is Hermitian and $T\tilde{g} = \tilde{g}$ for each $\tilde{g} \in L^2 \cap \pi(\mathscr{A})$ we deduce

$$\int P_{\mathscr{A}} f \cdot g \, d\mu = \int Tf \cdot g \, d\mu = \int f \cdot Tg \, d\mu = \int f \cdot g \, d\mu$$

for all $\tilde{f} \in L^2 \cap L^\infty$ and $\tilde{g} \in L^2 \cap \pi(\mathscr{A})$. Thus $P_{\mathscr{A}}$ satisfies condition (*).

Assume now that $Q: L^2 \cap L^\infty \to \overline{L^2 \cap \pi(\mathscr{A})}$ is another linear mapping satisfying condition (*). Let us note that for each $\tilde{f} \in L^2 \cap L^\infty$, $P_{\mathscr{A}} f$ and Qf are $\mathscr{F}$-measurable (see Remark 1) at the end of section 1). By condition (*) we have for every $F \in \mathscr{F}_0$:

$$\int P_{\mathscr{A}} f \cdot \varphi_F \, d\mu = \int Qf \cdot \varphi_F \, d\mu.$$

Hence (see Remark 3) at the end of section 1) $P_{\mathscr{A}} \tilde{f} = Q\tilde{f}$.

To prove 1), let $\tilde{h} \in L^2 \cap L^\infty$, $\tilde{h} \geqslant \tilde{0}$. Then $P_{\mathscr{A}} h$ is $\mathscr{F}$-measurable and for every $F \in \mathscr{F}_0$ we have:

$$\int P_{\mathscr{A}} h \cdot \varphi_F \, d\mu = \int h \cdot \varphi_F \, d\mu \geqslant 0.$$

It follows (see Remark 3) at the end of section 1) that $P_{\mathscr{A}} \tilde{h} \geqslant \tilde{0}$.

To prove 2), let $\tilde{f} \in L^2 \cap L^\infty$, $\tilde{f} \neq \tilde{0}$ and let $a = N_\infty(\tilde{f})$. We have $-a \leqslant f \leqslant a$ almost everywhere and hence for every $F \in \mathscr{F}_0$

$$-a\,\mu(F) \leqslant \int P_{\mathscr{A}} f \cdot \varphi_F \, d\mu = \int f \cdot \varphi_F \, d\mu \leqslant a\,\mu(F).$$

It follows (use Remark 3) at the end of section 1) that $-a \leqslant P_{\mathscr{A}} f \leqslant a$ almost everywhere.

3) We know that $P_{\mathscr{A}}(L^2 \cap L^\infty) \subset \overline{L^2 \cap \pi(\mathscr{A})}$ and by 2) above that $P_{\mathscr{A}}(L^2 \cap L^\infty) \subset L^\infty$. We deduce (see Proposition 3, section 1) that

$$P_{\mathscr{A}}(L^2 \cap L^\infty) \subset \overline{L^2 \cap \pi(\mathscr{A})} \cap L^\infty = L^2 \cap \pi(\mathscr{A}).$$

Since the converse inclusion is obvious, the equality $P_{\mathscr{A}}(L^2 \cap L^\infty) = L^2 \cap \pi(\mathscr{A})$ is established. Clearly $P_{\mathscr{A}}^2 = P_{\mathscr{A}}$.

4) Let $\tilde{f}\in L^2\cap L^\infty$ and $\tilde{g}\in L^2\cap\pi(\mathscr{A})$. By condition (*) we have for every $F\in\mathscr{F}_0$:

$$\int P_{\mathscr{A}}(fg)\cdot\varphi_F\,d\mu=\int(fg)\cdot\varphi_F\,d\mu=\int f\cdot(g\varphi_F)\,d\mu=\int(P_{\mathscr{A}}f)g\cdot\varphi_F\,d\mu$$

and therefore (use Remark 3) at the end of section 1) $P_{\mathscr{A}}(\tilde{g}\tilde{f})=\tilde{g}P_{\mathscr{A}}\tilde{f}$. This completes the proof of the proposition.

Note that if we take $\mathscr{C}=\mathscr{F}_0$ and $E=L^2\cap L^\infty$, the conditions introduced at the beginning of section 3 are satisfied and the linear mapping $T=P_{\mathscr{A}}$ commutes with every $\tilde{\varphi}_F$, $F\in\mathscr{F}_0$. We may then give the following definition (see Proposition 5 and the note following it, in section 3):

Definition 2. – *We call projection onto $\mathscr{A}$ (conditional expectation corresponding to $\mathscr{A}$) the canonical extension to L^∞ of the linear mapping $P_{\mathscr{A}}$ given by Proposition* 7.

We next give several useful properties of the projection onto $\mathscr{A}$ as well as a criterion for uniqueness:

Theorem 2.– *Let $P_{\mathscr{A}}$ be the projection onto $\mathscr{A}$. Then*

1) $P_{\mathscr{A}}: L^\infty\to L^\infty$ *is a positive linear mapping and* $P_{\mathscr{A}}\tilde{1}=\tilde{1}$.

2) $P_{\mathscr{A}}$ *maps* L^∞ *onto* $\pi(\mathscr{A})$ *and* $P^2_{\mathscr{A}}=P_{\mathscr{A}}$ *(in particular* $P_{\mathscr{A}}\tilde{f}=\tilde{f}$ *for every* $\tilde{f}\in\pi(\mathscr{A})$).

3) $P_{\mathscr{A}}(\tilde{g}\tilde{f})=\tilde{g}P_{\mathscr{A}}\tilde{f}$ *for all* $\tilde{f}\in L^\infty$ *and* $\tilde{g}\in\pi(\mathscr{A})$.

4) $P_{\mathscr{A}}|L^1\cap L^\infty$ *is a Dunford-Schwartz operator.*

Moreover the linear mapping $P_{\mathscr{A}}: L^\infty\to L^\infty$ *is uniquely determined by conditions* 1), 2), 3), *and* 4).

Proof: 1) The fact that $P_{\mathscr{A}}: L^\infty\to L^\infty$ is a positive linear mapping follows from Proposition 7 and Proposition 5. To see that $P_{\mathscr{A}}\tilde{1}=\tilde{1}$ we only need to note that for every $F\in\mathscr{F}_0$ we have:

$$\tilde{\varphi}_F P_{\mathscr{A}}\tilde{1}=P_{\mathscr{A}}(\tilde{\varphi}_F\tilde{1})=P_{\mathscr{A}}(\tilde{\varphi}_F)=\tilde{\varphi}_F.$$

2) Let $\tilde{f}\in L^\infty$. For each $F\in\mathscr{F}_0$ we have (use 3) of Proposition 7):

$$\tilde{\varphi}_F P_{\mathscr{A}}\tilde{f}=P_{\mathscr{A}}(\tilde{\varphi}_F\tilde{f})\in\pi(\mathscr{A}).$$

Since F is arbitrary and $\mathscr{A}$ is admissible we deduce $P_{\mathscr{A}}\tilde{f}\in\pi(\mathscr{A})$. Hence $P_{\mathscr{A}}(L^\infty)\subset\pi(\mathscr{A})$. Conversely let $\tilde{g}\in\pi(\mathscr{A})$. For each $F\in\mathscr{F}_0$, $\tilde{\varphi}_F\tilde{g}\in\pi(\mathscr{A})\cap L^2$ and hence (use 3) of Proposition 7):

$$\tilde{\varphi}_F P_{\mathscr{A}}\tilde{g}=P_{\mathscr{A}}(\tilde{\varphi}_F\tilde{g})=\tilde{\varphi}_F\tilde{g}.$$

We deduce $\tilde{g}=P_{\mathscr{A}}\tilde{g}\in P_{\mathscr{A}}(L^\infty)$; hence $\pi(\mathscr{A})\subset P_{\mathscr{A}}(L^\infty)$. Therefore 2) is proved.

3) Let $\tilde{f} \in L^\infty$ and $\tilde{g} \in \pi(\mathscr{A})$. For every $F \in \mathscr{F}_0$ we have

$$\tilde{\varphi}_F P_{\mathscr{A}}(\tilde{g}\tilde{f}) = P_{\mathscr{A}}(\tilde{\varphi}_F(\tilde{g}\tilde{f})) = P_{\mathscr{A}}((\tilde{\varphi}_F\tilde{g})(\tilde{\varphi}_F\tilde{f})) = \tilde{\varphi}_F\tilde{g}P_{\mathscr{A}}(\tilde{\varphi}_F\tilde{f}) = \tilde{\varphi}_F\tilde{g}P_{\mathscr{A}}\tilde{f}.$$

Since F is arbitrary, 3) is proved.

4) From 1) above it follows easily that $P_{\mathscr{A}}|L^1 \cap L^\infty$ is a contraction for the L^∞-norm. Let us show that it is also a contraction for the L^1-norm. Let $\tilde{f} \in L^1 \cap L^\infty (\subset L^2)$. Let $g \in \mathscr{A}$ be a simple integrable function. Then $\tilde{g} \in L^2 \cap \pi(\mathscr{A})$ and by condition (*) in Proposition 7 we have

$$|\int P_{\mathscr{A}} f \cdot g\, d\mu| = |\int f \cdot g\, d\mu| \leqslant N_\infty(\tilde{g}) N_1(\tilde{f}).$$

Since $g \in \mathscr{A}$ was an arbitrary simple integrable function, it follows that (see Remark 4) at the end of section 1):

$$N_1(P_{\mathscr{A}}\tilde{f}) \leqslant N_1(\tilde{f}).$$

Thus 4) is also proved.

It remains to prove the uniqueness assertion. Let $Q\colon L^\infty \to L^\infty$ be a linear mapping satisfying 1), 2), 3), and 4). Let $\tilde{f} \in L^1 \cap L^\infty$, $\tilde{f} \geqslant \tilde{0}$; then $\tilde{f} \in L^2 \cap L^\infty$. Let now $F \in \mathscr{F}_0$. Using the fact that $Q(\tilde{\varphi}_F\tilde{f}) = \tilde{\varphi}_F Q\tilde{f}$, that Q is positive, that $\|Q\|_1 \leqslant 1$ and condition (*) in Proposition 7 we deduce:

$$\int Qf \cdot \varphi_F\, d\mu = \int Q(f\varphi_F)\, d\mu \leqslant \int f \cdot \varphi_F\, d\mu = \int P_{\mathscr{A}} f \cdot \varphi_F\, d\mu.$$

Since Qf and $P_{\mathscr{A}}f$ belong to $\mathscr{A}$ and since $F \in \mathscr{F}_0$ was arbitrary, we deduce that $Q\tilde{f} \leqslant P_{\mathscr{A}}\tilde{f}$. Since this is true for every $\tilde{f} \in L^1 \cap L^\infty$ with $\tilde{f} \geqslant \tilde{0}$ it follows that $Q \leqslant P_{\mathscr{A}}$ (that is $Q\tilde{f} \leqslant P_{\mathscr{A}}\tilde{f}$ for every $\tilde{f} \in L^\infty$, $\tilde{f} \geqslant \tilde{0}$). We shall show that this implies $Q = P_{\mathscr{A}}$. By linearity it is enough to show that $Q\tilde{f} = P_{\mathscr{A}}\tilde{f}$ for $\tilde{f} \in L^\infty$ satisfying $\tilde{0} \leqslant \tilde{f} \leqslant \tilde{1}$. Let then $\tilde{f} \in L^\infty$, $\tilde{0} \leqslant \tilde{f} \leqslant \tilde{1}$. The relations

$$Q\tilde{f} \leqslant P_{\mathscr{A}}\tilde{f}, \quad Q(\tilde{1} - \tilde{f}) \leqslant P_{\mathscr{A}}(\tilde{1} - \tilde{f})$$

and

$$\tilde{1} = Q\tilde{f} + Q(\tilde{1} - \tilde{f}), \quad \tilde{1} = P_{\mathscr{A}}\tilde{f} + P_{\mathscr{A}}(\tilde{1} - \tilde{f})$$

imply that $Q\tilde{f} = P_{\mathscr{A}}\tilde{f}$. This completes the proof of the theorem.

5. Increasing sequences of projections corresponding to admissible subalgebras

In this section we assume that $\mathscr{C}$ is a *fixed collection of non-negligible integrable sets such that* $\sup\{\tilde{K} | K \in \mathscr{C}\} = \tilde{X}$.

Before stating and proving the convergence theorem for increasing sequences of projections corresponding to admissible subalgebras we wish to make the following:

Remark. – Let $\mathscr{E} \subset M^{\infty}$ be a set containing $\{\varphi_K | K \in \mathscr{C}\}$. The intersection of all admissible subalgebras of M^{∞} containing $\mathscr{E}$ is an admissible subalgebra; we shall call this algebra the admissible subalgebra generated by $\mathscr{E}$.

Proof: It is obvious that the intersection of all admissible subalgebras of M^{∞} containing $\mathscr{E}$ is an algebra which contains 1 and $\mathscr{N}^{\infty}$ and is closed under pointwise convergence of bounded sequences (condition α) of Proposition 3). Using Remark 5) at the end of section 1 it is easily seen that this algebra satisfies also condition β) of Proposition 3 and hence is an admissble algebra.

Theorem 3. (The martingale convergence theorem)[1]. – *Let* $(\mathscr{A}_n)$ *be an increasing sequence of admissible subalgebras of* M^{∞} *and suppose that* $\mathscr{A}_1 \supset \{\varphi_K | K \in \mathscr{C}\}$. *For each n let* $P_n = P_{\mathscr{A}_n}$ *(= the projection onto* $\mathscr{A}_n$*). There is then a projection* $P_{\infty} \in \mathscr{D}$ *such that for each* $f \in \mathscr{V}$[2]*, the sequence* $(P_n f(x))$ *converges to* $P_{\infty} f(x)$ *almost everywhere. Moreover:*

1) P_{∞} *is positive and* P_{∞} *commutes with every* $\tilde{\varphi}_K$, $K \in \mathscr{C}$.

2) *For every* $\tilde{f} \in L^{\infty}$, *the sequence* $(P_n f(x))$ *converges to* $P_{\infty} f(x)$ *almost everywhere*[3].

3) *Let* $\mathscr{A}_{\infty} = \{f \in M^{\infty} | \tilde{f} \in P_{\infty}(L^{\infty})\}$. *Then* $\mathscr{A}_{\infty}$ *is the admissible algebra generated by* $\bigcup_n \mathscr{A}_n$ *and* P_{∞} *is the projection onto* $\mathscr{A}_{\infty}$.

Proof: 1) The existence of the projection $P_{\infty} \in \mathscr{D}$ such that for each $f \in \mathscr{V}$ the sequence $(P_n f(x))$ converges to $P_{\infty} f(x)$ almost everywhere is given by Theorem 3 in Appendix I.

Since each P_n is positive and commutes with every $\tilde{\varphi}_K$, $K \in \mathscr{C}$ (see Theorem 2, section 4) statement 1) follows.

2) Let $\tilde{f} \in L^{\infty}$. For each $K \in \mathscr{C}$, $(P_n(\varphi_K f))$ converges to $P_{\infty}(\varphi_K f)$ almost everywhere. Since $P_n(\varphi_K f) = \varphi_K P_n f$ for all n and $P_{\infty}(\varphi_K f) = \varphi_K P_{\infty} f$, this means that $(P_n f)$ converges to $P_{\infty} f$ almost everywhere on K. Since $K \in \mathscr{C}$ was arbitrary we deduce that $(P_n f)$ converges almost everywhere to $P_{\infty} f$.

We divide the proof of statement 3) into four parts:

a) We show first that

$$(\ast) \qquad P_{\infty}(\tilde{g}\tilde{f}) = \tilde{g} P_{\infty} \tilde{f}$$

for all $\tilde{f} \in L^{\infty}$ and $\tilde{g} \in P_{\infty}(L^{\infty})$; this obviously implies that $\mathscr{A}_{\infty}$ is an algebra. We first prove $(\ast)$ in the case when $\tilde{g} \in P_j(L^{\infty})$ for some integer j.

---

[1]) See [108]; see also [31], [82], and [96].

[2]) We use the notation $\mathscr{V} = \bigcup_{1 \leq p < +\infty} \mathscr{L}^p$ (see also Appendix I).

[3]) Here of course P_{∞} denotes the canonical extension to L^{∞}.

Since the sequence (P_n) is increasing we have for $n \geq j$, $P_n(\tilde{g}\tilde{f}) = \tilde{g}P_n\tilde{f}$ (see Theorem 2, section 4). By 2) above, the sequences $(P_n(gf))_{n\geq j}$ and $(gP_nf)_{n\geq j}$ converge almost everywhere to $P_\infty(gf)$ and $gP_\infty f$ respectively. Hence equation (⁑) is proved in this case. Let now $\tilde{g}$ be an arbitrary element in $P_\infty(L^\infty)$. By 2) above, we may assume that g is the (almost everywhere) pointwise limit of the bounded sequence (g_n), where $g_n = P_n g$ for each n. By what we just proved.

$$P_\infty(\tilde{g}_n\tilde{f}) = \tilde{g}_n P_\infty\tilde{f} \quad \text{for every } n.$$

Letting $n \to \infty$ and using Proposition 6, section 3, we deduce $P_\infty(\tilde{g}\tilde{f}) = \tilde{g}P_\infty\tilde{f}$. Thus equation (⁑) is proved and hence $\mathscr{A}_\infty$ is an algebra.

b) We next show that $\mathscr{A}_\infty$ is an admissible subalgebra of M^∞. Clearly $\mathscr{A}_\infty$ contains 1 and $\mathscr{N}^\infty$. Using Proposition 6, section 3 and the fact that P_∞ belongs to $\mathscr{D}_+$, we see that $\mathscr{A}_\infty$ is closed under pointwise convergence of bounded sequences (condition α) of Proposition 3). Let us show that $\mathscr{A}_\infty$ satisfies also condition β) of Proposition 3. Let $f: X \to R$ be a bounded function such that $f\varphi_F \in \mathscr{A}_\infty$ for every integrable set F with $\varphi_F \in \mathscr{A}_\infty$. In particular, $f\varphi_K \in \mathscr{A}_\infty$ for every $K \in \mathscr{C}$. It follows that f is measurable and that $f\varphi_K = P_\infty(f\varphi_K)$ for every $K \in \mathscr{C}$. By 1) above, $P_\infty(f\varphi_K) = \varphi_K P_\infty f$ for each $K \in \mathscr{C}$. We deduce that $f \equiv P_\infty f$ and thus $f \in \mathscr{A}_\infty$. Hence the statement that $\mathscr{A}_\infty$ is an admissible algebra is proved.

c) We now show that $\mathscr{A}_\infty$ is the admissible algebra generated by $\bigcup_n \mathscr{A}_n$. Clearly $\mathscr{A}_\infty \supset \bigcup_n \mathscr{A}_n$. Let $f \in \mathscr{A}_\infty$. By 2) above, f is the almost everywhere limit of the bounded sequence $(P_n f)$. Since $(P_n f)$ is a sequence in $\bigcup_n \mathscr{A}_n$, we deduce that f belongs to any admissible algebra containing $\bigcup_n \mathscr{A}_n$. Therefore the assertion is proved.

d) It remains to show that P_∞ is the projection onto $\mathscr{A}_\infty$. Since P_∞ satisfies conditions 1), 2), 3) and 4) of Theorem 2, section 4 (make use of equation (⁑) given in the proof of a) above) the statement follows from the uniqueness assertion of Theorem 2. This completes the proof of the theorem.

CHAPTER III

Basic definitions and remarks concerning the notion of lifting

In this chapter we introduce the notions of lifting, linear lifting and lower density for an admissible subalgebra of M^∞ and we give several results concerning these notions.

We assume throughout this chapter that $(X, N, \mathscr{R})$ *is localizable, that* $\mathscr{A}$ *is an admissible subalgebra of* M^∞ *and* $\mathscr{F} = \{A \mid \varphi_A \in \mathscr{A}\}$ *(we recall that* $\mathscr{F}$ *is a tribe of subsets of* X*).*

1. Linear liftings and liftings of an admissible subalgebra. Lower densities

Let ρ be a mapping of $\mathscr{A}$ into $\mathscr{A} (\subset M^\infty)$ and consider the following conditions:

(I) $\rho(f) \equiv f$;
(II) $f \equiv g$ implies $\rho(f) = \rho(g)$;
(III) $\rho(1) = 1$;
(IV) $f \geqslant 0$ implies $\rho(f) \geqslant 0$;
(V) $\rho(af + bg) = a\rho(f) + b\rho(g)$;
(VI) $\rho(fg) = \rho(f)\rho(g)$.

Note that (VI) $\Rightarrow$ (IV).

Definition 1. – *A mapping* $\rho: \mathscr{A} \to \mathscr{A}$ *is a linear lifting of* $\mathscr{A}$ *if it satisfies* (I)–(V).

Definition 2. – *A mapping* $\rho: \mathscr{A} \to \mathscr{A}$ *is a lifting of* $\mathscr{A}$ *if it satisfies* (I)–(VI).

Let us remark that if ρ is a linear lifting of $\mathscr{A}$, we may define $\rho(\tilde{f})$ unambiguously by the equations $\rho(\tilde{f}) = \rho(f)$ for $\tilde{f} \in \pi(\mathscr{A})$. In particular if ρ is a lifting of $\mathscr{A}$ this yields a *cross-section* of the canonical mapping $\pi: \mathscr{A} \to \mathscr{A}/\mathscr{N}^\infty$ *preserving the algebraic operations.*

Note that if ρ is a *linear lifting* of $\mathscr{A}$, then ρ has the following "continuity property"[1]):

(1) $$\|\rho(f)\|_\infty = N_\infty(f) \leqslant \|f\|_\infty \quad \text{for any} \quad f \in \mathscr{A}.$$

This follows from the inequalities

$$-N_\infty(f) \leqslant f(x) \leqslant N_\infty(f)$$

which hold almost everywhere.

Note also that if ρ is a *lifting* of $\mathscr{A}$ then for each $h \in C_R(R)$

(2) $$\rho(h \circ f) = h \circ \rho(f) \quad \text{for} \quad f \in \mathscr{A}.$$

In fact, it is enough to note that (2) holds whenever h is a polynomial function, to use the Stone-Weierstrass theorem and the "continuity property" (1) of ρ. In particular we have:

(2') $$\rho(|f|) = |\rho(f)| \quad \text{for} \quad f \in \mathscr{A}.$$

Using (2') and the linearity of ρ we deduce

(3) $$\rho(\sup(f,g)) = \sup(\rho(f), \rho(g)), \qquad \rho(\inf(f,g)) = \inf(\rho(f), \rho(g))$$

for any $f \in \mathscr{A}$, $g \in \mathscr{A}$.

There is an *equivalent* way of defining a lifting if one prefers to work with sets rather than functions. We recall that for $A \in \mathscr{B}$, $B \in \mathscr{B}$ we write $A \equiv B$ if $A \triangle B = (A \cap \complement B) \cup (B \cap \complement A) \in \mathscr{N}_0$. We recall that a bounded function $f: X \to R$ belongs to $\mathscr{A}$ if and only if f is $\mathscr{F}$-measurable. Let now ρ be a lifting of $\mathscr{A}$. For each $A \in \mathscr{F}$, $\rho(\varphi_A)$ is the characteristic function of a uniquely determined set belonging to $\mathscr{F}$ (use (VI)). We denote this set by $\rho(A)$; thus

(4) $$\varphi_{\rho(A)} = \rho(\varphi_A).$$

The mapping $\rho: \mathscr{F} \to \mathscr{F}$ thus obtained satisfies the conditions:

(I') $\rho(A) \equiv A$;
(II') $A \equiv B$ implies $\rho(A) = \rho(B)$;
(III') $\rho(X) = X$, $\rho(\varnothing) = \varnothing$;
(IV') $\rho(A \cap B) = \rho(A) \cap \rho(B)$;
(V') $\rho(A \cup B) = \rho(A) \cup \rho(B)$.

This leads to the following

---

[1]) We recall that for a bounded function $f: X \to R$,

$$\|f\|_\infty = \sup_{x \in X} |f(x)|.$$

Definition 3. – *A mapping* $\rho: \mathscr{F} \to \mathscr{F}$ *is a lifting of* $\mathscr{F}$ *if it satisfies* (I′)–(V′).

As we saw above, given a lifting ρ of $\mathscr{A}$ we define a lifting of $\mathscr{F}$ by (4).

We shall use the same letter to denote the lifting of $\mathscr{A}$ and the corresponding lifting of $\mathscr{F}$.

Conversely, given a lifting ρ of $\mathscr{F}$ there exists a unique lifting ρ of $\mathscr{A}$ satisfying (4). This can be proved as follows (see [65]):

Denote by $\mathscr{S}(\mathscr{F})$ the algebra of all simple functions which are "based" on sets belonging to $\mathscr{F}$. For $f = \sum_{j=1}^{q} c_j \varphi_{A_j} \in \mathscr{S}(\mathscr{F})$ (here $A_1 \in \mathscr{F}, \ldots, A_q \in \mathscr{F}$) define $\rho(f) = \sum_{j=1}^{q} c_j \varphi_{\rho(A_j)}$. It is easy to see that $\rho(f)$ is well defined and that $\|\rho(f)\|_\infty = N_\infty(f)$ for each $f \in \mathscr{S}(\mathscr{F})$; the properties (I)–(VI) are also satisfied by ρ on $\mathscr{S}(\mathscr{F})$. Since $\mathscr{S}(\mathscr{F})$ is dense in $\mathscr{A}$ (for the topology defined by N_∞), ρ can be extended by continuity to $\mathscr{A}$ and the extension continues to satisfy conditions (I)–(VI). It is also clear that the equations (4) are satisfied. To finish the proof note that if ρ_1 and ρ_2 are two liftings of $\mathscr{A}$ such that

$$\rho_1(\varphi_A) = \rho_2(\varphi_A) \quad \text{for} \quad A \in \mathscr{F},$$

then ρ_1 and ρ_2 coincide on $\mathscr{S}(\mathscr{F})$ and hence on $\mathscr{A}$.

Definition 4. – *A mapping* $\delta: \mathscr{F} \to \mathscr{F}$ *is a lower density of* $\mathscr{F}$ *if it satisfies* (I′)–(IV′).

Definition 5. – *A mapping* $\delta: \mathscr{F} \to \mathscr{F}$ *is an upper density of* $\mathscr{F}$ *if it satisfies* (I′)–(III′) *and* (V′).

Let us note that if $\delta: \mathscr{F} \to \mathscr{F}$ is a lower density of $\mathscr{F}$ and if we define $\delta': \mathscr{F} \to \mathscr{F}$ by

$$\delta'(A) = \complement\, \delta(\complement A), \quad \text{for} \quad A \in \mathscr{F}$$

then δ' is an upper density of $\mathscr{F}$.

Let now $\gamma: \mathscr{A} \to \mathscr{A}$ be a *linear lifting* of $\mathscr{A}$. For each $A \in \mathscr{F}$ define

$$\theta(A) = \{x \mid \gamma(\varphi_A)(x) = 1\}.$$

It is clear that $\theta: \mathscr{F} \to \mathscr{F}$ and it is easily seen that θ is in fact a *lower density* of $\mathscr{F}$. The conditions (I′), (II′), (III′) are obviously satisfied. It remains to check (IV′). Let $A \in \mathscr{F}$, $B \in \mathscr{F}$ and let $x \in \theta(A) \cap \theta(B)$. Since for any $E \in \mathscr{F}$, $\gamma(\varphi_{\complement E}) = 1 - \gamma(\varphi_E)$, we have

$$\gamma(\varphi_A)(x) = 1, \quad \gamma(\varphi_B)(x) = 1 \;\Rightarrow\; \gamma(\varphi_{\complement A})(x) = 0, \quad \gamma(\varphi_{\complement B})(x) = 0.$$

Since $\varphi_{CA\cup CB}\leqslant\varphi_{CA}+\varphi_{CB}$ we deduce

$$\gamma(\varphi_{CA\cup CB})(x)=0 \Rightarrow \gamma(\varphi_{C(A\cap B)})(x)=0 \Rightarrow \gamma(\varphi_{A\cap B})(x)=1 \Rightarrow x\in\theta(A\cap B).$$

Thus $\theta(A)\cap\theta(B)\subset\theta(A\cap B)$. Since the relations $E\in\mathscr{F}$, $F\in\mathscr{F}$, $E\subset F$ imply $\theta(E)\subset\theta(F)$ (see the definition of θ), we also have the converse inclusion $\theta(A\cap B)\subset\theta(A)\cap\theta(B)$ and (IV′) is verified. Hence θ is a lower density of $\mathscr{F}$; θ is called *the lower density associated with the linear lifting* γ *of* $\mathscr{A}$. For each $A\in\mathscr{F}$ define also

$$\theta'(A)=\{x\,|\,\gamma(\varphi_A)(x)>0\}.$$

It is immediate that $\theta'(A)=\complement\theta(\complement A)$ for each $A\in\mathscr{F}$; θ' is called *the upper density associated with the linear lifting* γ *of* $\mathscr{A}$.

2. Linear liftings, liftings and extremal points

We shall denote by $\mathscr{G}_{\mathscr{A}}$ the set of all *linear liftings* of $\mathscr{A}$. We shall also introduce the:

Definition 6. – *Let* $\gamma\in\mathscr{G}_{\mathscr{A}}$ *and let* θ *and* θ' *be the lower and upper densities associated with* γ. *We denote by* $\mathscr{G}_{\mathscr{A}}(\gamma)$ *the set of all* $\sigma\in\mathscr{G}_{\mathscr{A}}$ *satisfying the inequalities*

$$\varphi_{\theta(A)}\leqslant\sigma(\varphi_A)\leqslant\varphi_{\theta'(A)}$$

for every $A\in\mathscr{F}$.

Consider the (convex) set $\mathscr{L}_1^*(\pi(\mathscr{A}),\mathscr{A})$ (see section 2, chapter 2). For every $\sigma\in\mathscr{G}_{\mathscr{A}}$ define the mapping $\tilde{\sigma}:\pi(\mathscr{A})\to\mathscr{A}$ by the equations

$$\tilde{\sigma}(\tilde{f})=\sigma(f)$$

for $f\in\mathscr{A}$. Clearly $\tilde{\sigma}$ is well defined, $\tilde{\sigma}$ belongs to $\mathscr{L}_1^*(\pi(\mathscr{A}),\mathscr{A})$ and

$$\tilde{\sigma}(\tilde{f})\equiv f$$

for every $f\in\mathscr{A}$. Conversely if $\delta\in\mathscr{L}_1^*(\pi(\mathscr{A}),\mathscr{A})$ is such that

$$\delta(\tilde{f})\equiv f \tag{1}$$

for every $f\in\mathscr{A}$ and if we define $\sigma:\mathscr{A}\to\mathscr{A}$ by

$$\sigma(f)=\delta(\tilde{f})$$

for $f\in\mathscr{A}$ then $\sigma\in\mathscr{G}_{\mathscr{A}}$ and $\tilde{\sigma}=\delta$. It follows that $\sigma\to\tilde{\sigma}$ identifies $\mathscr{G}_{\mathscr{A}}$ with the (convex) subset of $\mathscr{L}_1^*(\pi(\mathscr{A}),\mathscr{A})$ consisting of all $\delta\in\mathscr{L}_1^*(\pi(\mathscr{A}),\mathscr{A})$ satisfying (1). Note also that if $\delta\in\mathscr{L}_1^*(\pi(\mathscr{A}),\mathscr{A})$ satisfies (1) then

$$\|\delta(\tilde{f})\|_\infty\leqslant N_\infty(\tilde{f})\leqslant\|f\|_\infty \tag{2}$$

for every $f\in\mathscr{A}$.

We denote now by R^I the locally convex space

$$\prod_{(f,x)\in I} R^{(f,x)}$$

where $I=\mathscr{A}\times X$ and $R^{(f,x)}=R$ (= the real line with the usual topology) for all $(f,x)\in I$. Every $\sigma\in\mathscr{G}_{\mathscr{A}}$ can be identified with the element $(\sigma(f)(x))_{(f,x)\in I}$ of R^I, hence $\mathscr{G}_{\mathscr{A}}$ can be identified with a part of R^I; it is easy to see that this set is *convex*.

Theorem 1. – *Let* $\gamma\in\mathscr{G}_{\mathscr{A}}$. *Then:*

1) *The set* $\mathscr{G}_{\mathscr{A}}(\gamma)$ *is convex and compact in* R^I.

2) *An element* $\sigma\in\mathscr{G}_{\mathscr{A}}(\gamma)$ *is extremal in* $\mathscr{G}_{\mathscr{A}}(\gamma)$ *if and only if* σ *is multiplicative (i.e. a lifting of* $\mathscr{A}$*).*

Proof: 1) It is obvious that $\mathscr{G}_{\mathscr{A}}(\gamma)$ is convex. On the other hand for $\sigma\in\mathscr{G}_{\mathscr{A}}(\gamma)$ and $(f,x)\in I$ we have $|\sigma(f)(x)|\leqslant N_\infty(f)$; hence $\mathscr{G}_{\mathscr{A}}(\gamma)$ is a *bounded part* of R^I. We shall show that $\mathscr{G}_{\mathscr{A}}(\gamma)$ is also *closed* in R^I. Recall that we denoted by $B^\infty(X)$ the Banach algebra of all bounded real-valued functions on X (endowed with the supremum norm $f\to\|f\|_\infty$). Let now $(\sigma_j)_{j\in J}$ be a directed family of elements of $\mathscr{G}_{\mathscr{A}}(\gamma)$ converging to some element of R^I. For each $f\in\mathscr{A}$ and $x\in X$ we write

$$\lim_{j\in J}\sigma_j(f)(x)=\sigma_\infty(f)(x).$$

It is clear that σ_∞ as a mapping of $\mathscr{A}$ into B^∞ satisfies the conditions (II)–(V). In particular σ_∞ has the "continuity property"

$$\|\sigma_\infty(f)\|_\infty\leqslant N_\infty(f)\leqslant\|f\|_\infty \quad \text{for each} \quad f\in\mathscr{A}.$$

We also have

$$\varphi_{\theta(A)}\leqslant\sigma_\infty(\varphi_A)\leqslant\varphi_{\theta'(A)} \quad \text{for every} \quad A\in\mathscr{F}.$$

The previous relations show that $\sigma_\infty(\varphi_A)\in\mathscr{A}$ and $\sigma_\infty(\varphi_A)\equiv\varphi_A$ for each $A\in\mathscr{F}$. Hence $\sigma_\infty(f)\in\mathscr{A}$ and $\sigma_\infty(f)\equiv f$ for every $f\in\mathscr{S}(\mathscr{F})$ (recall that $\mathscr{S}(\mathscr{F})$ is the set of simple functions "based" on sets belonging to $\mathscr{F}$). Since σ_∞ is continuous as a mapping of $\mathscr{A}$ into $B^\infty(X)$ (both endowed with the supremum norm) and since $\mathscr{S}(\mathscr{F})$ is dense in $\mathscr{A}$, it follows that $\sigma_\infty(f)\in\mathscr{A}$ and $\sigma_\infty(f)\equiv f$ for all $f\in\mathscr{A}$. Thus $\sigma_\infty\in\mathscr{G}_{\mathscr{A}}(\gamma)$ and $\mathscr{G}_{\mathscr{A}}(\gamma)$ is closed. Hence statement 1) is proved.

2) Since the (positive) characters of $\mathscr{A}$ clearly separate the elements of $\mathscr{A}$, it follows that (see the Corollary at the end of section 2, chapter 2) an element in $\mathscr{L}_1^*(\pi(\mathscr{A}),\mathscr{A})$ is extremal if and only if it is multiplicative. On the other hand we will prove in Proposition 1 below that an element $\sigma\in\mathscr{G}_{\mathscr{A}}(\gamma)$ is extremal in $\mathscr{G}_{\mathscr{A}}(\gamma)$ if and only if it is extremal in $\mathscr{L}_1^*(\pi(\mathscr{A}),\mathscr{A})$. This will complete the proof of statement 2).

Hence Theorem 1 is completely proved.

Proposition 1. – *Let $\sigma \in \mathcal{G}_{\mathcal{A}}(\gamma)$. Then σ is extremal in $\mathcal{G}_{\mathcal{A}}(\gamma)$ if and only if σ is extremal in $\mathcal{L}_1^*(\pi(\mathcal{A}),\mathcal{A})$.*

Proof: Since $\mathcal{G}_{\mathcal{A}}(\gamma) \subset \mathcal{L}_1^*(\pi(\mathcal{A}),\mathcal{A})$ it is clear that if σ is extremal in $\mathcal{L}_1^*(\pi(\mathcal{A}),\mathcal{A})$ then σ is extremal in $\mathcal{G}_{\mathcal{A}}(\gamma)$.

Conversely suppose that σ is extremal in $\mathcal{G}_{\mathcal{A}}(\gamma)$ and that σ admits a representation of the form

$$\sigma = t\,T_1 + (1-t)\,T_2 \tag{3}$$

with T_1 and T_2 in $\mathcal{L}_1^*(\pi(\mathcal{A}),\mathcal{A})$ and $0<t<1$. We shall show that relation (3) implies $T_1 \in \mathcal{G}_{\mathcal{A}}(\gamma)$ and $T_2 \in \mathcal{G}_{\mathcal{A}}(\gamma)$; this obviously leads to $T_1 = T_2$ and hence proves that σ is extremal in $\mathcal{L}_1^*(\pi(\mathcal{A}),\mathcal{A})$.

Since $0 \leqslant T_\iota(\tilde{\varphi}_A) \leqslant 1$ for $A \in \mathcal{F}$ and $i=1,2$ and since (recall that $\sigma \in \mathcal{G}_{\mathcal{A}}(\gamma)$)

$$\varphi_{\theta(A)} \leqslant t\,T_1(\tilde{\varphi}_A) + (1-t)\,T_2(\tilde{\varphi}_A) \leqslant \varphi_{\theta'(A)}$$

for $A \in \mathcal{F}$ we deduce

$$\varphi_{\theta(A)} \leqslant T_\iota(\tilde{\varphi}_A) \leqslant \varphi_{\theta'(A)}$$

for $A \in \mathcal{F}$ and $i=1,2$. This shows that $T_\iota(\tilde{\varphi}_A) \equiv \varphi_A$ for each $A \in \mathcal{F}$ and $i=1,2$. Repeating the argument used in the last part of the proof of statement 1) of Theorem 1, we deduce that $T_\iota \in \mathcal{G}_{\mathcal{A}}(\gamma)$ $(i=1,2)$. Thus Proposition 1 is proved.

Theorem 2. – *Let $\gamma \in \mathcal{G}_{\mathcal{A}}$. Then the set $\mathcal{G}_{\mathcal{A}}(\gamma)$ contains a lifting ρ of $\mathcal{A}$. Moreover if $E \in \mathcal{F}$ and $\gamma(\varphi_E)$ is a characteristic function, then $\rho(\varphi_E) = \gamma(\varphi_E)$. In particular if $\mathcal{A}_0 \subset \mathcal{A}$ is an admissible subalgebra and if $\gamma|\mathcal{A}_0$ is a lifting of $\mathcal{A}_0$ then $\rho|\mathcal{A}_0 = \gamma|\mathcal{A}_0$.*

Proof: The existence of a lifting ρ of $\mathcal{A}$ in the set $\mathcal{G}_{\mathcal{A}}(\gamma)$ follows from Theorem 1 and the Krein-Milman theorem applied to the set $\mathcal{G}_{\mathcal{A}}(\gamma)$.

If $E \in \mathcal{F}$ and $\gamma(\varphi_E)$ is a characteristic function then clearly $\varphi_{\theta(A)} = \varphi_{\theta'(A)} = \gamma(\varphi_E)$ and hence $\rho(\varphi_E) = \gamma(\varphi_E)$.

Let now $\mathcal{A}_0 \subset \mathcal{A}$ be an admissible subalgebra and let $\mathcal{U} = \{A \mid \varphi_A \in \mathcal{A}_0\}$. Suppose that $\gamma|\mathcal{A}_0$ is a lifting of $\mathcal{A}_0$. Then $\gamma(\varphi_E)$ is a characteristic function for each $E \in \mathcal{U}$, whence $\gamma(\varphi_E) = \rho(\varphi_E)$ for all $E \in \mathcal{U}$. Hence $\gamma(f) = \rho(f)$ for every simple function f based on sets belonging to $\mathcal{U}$ and therefore (by approximation) $\gamma(f) = \rho(f)$ for every $f \in \mathcal{A}_0$.

A different proof of the existence of a lifting ρ of $\mathcal{A}$ belonging to the set $\mathcal{G}_{\mathcal{A}}(\gamma)$ will be given later, using the "density topology" $\mathcal{T}_\theta$.

Remarks. – 1) Denote by $\mathscr{G}_{\mathscr{A}}^*$ the set of all *liftings of* $\mathscr{A}$. We have

$$\mathscr{G}_{\mathscr{A}}^* = \textit{the set of all extremal elements of } \mathscr{G}_{\mathscr{A}}.$$

In fact if $\sigma \in \mathscr{G}_{\mathscr{A}}$ is extremal in $\mathscr{G}_{\mathscr{A}}$ then σ is extremal in $\mathscr{G}_{\mathscr{A}}(\sigma)$ (since $\sigma \in \mathscr{G}_{\mathscr{A}}(\sigma) \subset \mathscr{G}_{\mathscr{A}}$) and hence σ is extremal in $\mathscr{L}_1^*(\pi(\mathscr{A}),\mathscr{A})$. We conclude that σ is multiplicative, that is $\sigma \in \mathscr{G}_{\mathscr{A}}^*$.

Conversely if $\sigma \in \mathscr{G}_{\mathscr{A}}^*$ then σ is a multiplicative element of $\mathscr{L}_1^*(\pi(\mathscr{A}),\mathscr{A})$. Since $\mathscr{G}_{\mathscr{A}} \subset \mathscr{L}_1^*(\pi(\mathscr{A}),\mathscr{A})$ we conclude that σ is extremal in $\mathscr{G}_{\mathscr{A}}$.

2) Note also that by Theorems 1 and 2 we have

$$\mathscr{G}_{\mathscr{A}}^* \neq \varnothing \Leftrightarrow \mathscr{G}_{\mathscr{A}} \neq \varnothing.$$

We shall see in the next chapter that if $(X,N,\mathscr{R})$ is strictly localizable then

$$\mathscr{G}_{\mathscr{A}}^* \neq \varnothing$$

for every admissible subalgebra $\mathscr{A}$.

3. On the measurability of the upper envelope. A limit theorem

We assume in this section that ρ *is a lifting of the admissible algebra* $\mathscr{A}$. The result that we shall give here (see [67]) will be essentially used in the proof of the existence of a lifting. It is also of independent interest since for abstract measure spaces the usual limit theorems are valid only for sequences of functions.

Theorem 3. – *Let* $\mathscr{H} \subset \mathscr{A}_+$ *be a set directed for the relation* $\leqslant$ *and such that* $f \leqslant \rho(f)$ *for every* $f \in \mathscr{H}$. *Then the upper envelope* u *of* $\mathscr{H}$ *is* $\mathscr{F}$-*measurable and*

$$N(u) = \sup_{f \in \mathscr{H}} N(f).$$

Furthermore, if $\mathscr{H}$ *is bounded above, then* $u \in \mathscr{A}$, $u \leqslant \rho(u)$ *and* $\tilde{u} = \sup\{\tilde{f} \mid f \in \mathscr{H}\}$.

Proof: Let $\mathscr{F}_0 = \{F \mid F \in \mathscr{B}_0,\ \varphi_F \in \mathscr{A}\}$. We recall that the conditions α) and β) of Proposition 3 in section 1, chapter 2 are satisfied.

Consider first the case when $\mathscr{H}$ is bounded above by some constant $a \geqslant 0$. Let $F \in \mathscr{F}_0$ with $\mu(F) > 0$ and define

$$\mathscr{H}(F) = \{f\varphi_{\rho(F)} \mid f \in \mathscr{H}\}.$$

It is clear that the set $\mathscr{H}(F)$ is directed for the relation $\leqslant$, that $g \leqslant \rho(g)$ for each $g \in \mathscr{H}(F)$ and that $u\varphi_{\rho(F)}$ is the upper envelope of $\mathscr{H}(F)$.

Also $\mathscr{H}(F) \subset \mathscr{L}^1$ and

$$\alpha = \sup_{g \in \mathscr{H}(F)} N_1(g) < \infty .$$

Let (g_n) be an increasing sequence in $\mathscr{H}(F)$ such that

$$\alpha = \sup_n N_1(g_n).$$

Let g_∞ be the upper envelope of the sequence (g_n). Then (see Theorem 4 in section 2, chapter 1):

$$\tilde{g}_\infty = \sup \{\tilde{g} | g \in \mathscr{H}(F)\}.$$

It follows that

$$\tilde{g}_\infty \geqslant \tilde{g} \quad \text{for each} \quad g \in \mathscr{H}(F);$$

hence

$$\rho(g_\infty) \geqslant \rho(g) \geqslant g \quad \text{for each} \quad g \in \mathscr{H}(F)$$

whence

$$\rho(g_\infty) \geqslant u \varphi_{\rho(F)} \geqslant g_\infty .$$

The last relation shows that $u \varphi_{\rho(F)} \equiv g_\infty$ and hence proves that $u \varphi_{\rho(F)}$ is $\mathscr{F}$-measurable and that

(*) $$\tilde{u} \tilde{\varphi}_{\rho(F)} = \sup \{\tilde{f} \tilde{\varphi}_{\rho(F)} | f \in \mathscr{H}\}.$$

It follows that:

($\genfrac{}{}{0pt}{}{*}{*}$) $$N_1(u \varphi_{\rho(F)}) = \sup_{f \in \mathscr{H}} N_1(f \varphi_{\rho(F)}).$$

Since u is bounded and $u \varphi_{\rho(F)} \in \mathscr{A}$ for each $F \in \mathscr{F}_0$, property β) of an admissible subalgebra implies that $u \in \mathscr{A}$. Also since $u \geqslant f$ for each $f \in \mathscr{H}$, we deduce $\rho(u) \geqslant \rho(f) \geqslant f$ for each $f \in \mathscr{H}$ which shows that $\rho(u) \geqslant u$. Finally since the relations (*) and ($\genfrac{}{}{0pt}{}{*}{*}$) hold for every $F \in \mathscr{F}_0$ and since $\sup\{\tilde{F} | F \in \mathscr{F}_0\} = \tilde{X}$ we deduce (use Theorem 14, section 5, chapter 1 and Remark 2 at the end of section 1, chapter 2, respectively):

$$\tilde{u} = \sup_{F \in \mathscr{F}_0} \tilde{u} \tilde{\varphi}_F = \sup_{F \in \mathscr{F}_0} \left(\sup_{f \in \mathscr{H}} \tilde{f} \tilde{\varphi}_F \right) = \sup_{f \in \mathscr{H}} \left(\sup_{F \in \mathscr{F}_0} \tilde{f} \tilde{\varphi}_F \right) = \sup_{f \in \mathscr{H}} \tilde{f}$$

and

$$N_1(u) = \sup_{F \in \mathscr{F}_0} N_1(u \varphi_F) = \sup_{F \in \mathscr{F}_0} \left(\sup_{f \in \mathscr{H}} N_1(f \varphi_F) \right)$$

$$= \sup_{f \in \mathscr{H}} \left(\sup_{F \in \mathscr{F}_0} N_1(f \varphi_F) \right) = \sup_{f \in \mathscr{H}} N_1(f).$$

This completes the proof in the case when $\mathscr{H}$ is bounded above.

Consider now the general case. For each positive integer n define

$$\mathscr{H}_n = \{\inf(f, n) \mid f \in \mathscr{H}\}$$

and

$$u_n = \inf(u, n).$$

It is clear that $\mathscr{H}_n$ is directed for the relation $\leqslant$, that $g \leqslant \rho(g)$ for every $g \in \mathscr{H}_n$ (use relation (3) in section 1 of this chapter) and that u_n is the upper envelope of $\mathscr{H}_n$. Since $\mathscr{H}_n$ is bounded above by n, we deduce that $u_n \in \mathscr{A}$ and

$$N_1(u_n) = \sup_{f \in \mathscr{H}} N_1(\inf(f, n))$$

for each n. It follows that $u = \sup_n u_n$ is $\mathscr{F}$-measurable and

$$\begin{aligned} N_1(u) &= \sup_n N_1(u_n) = \sup_n \Big(\sup_{f \in \mathscr{H}} N_1(\inf(f, n))\Big) \\ &= \sup_{f \in \mathscr{H}} \Big(\sup_n N_1(\inf(f, n))\Big) = \sup_{f \in \mathscr{H}} N_1(f). \end{aligned}$$

This completes the proof of Theorem 3.

Remark. – In the case when $\mathscr{H}$ is bounded above and $N(1) < \infty$ (i.e. $X \in \mathscr{F}_0$), the first part of the proof shows that there is an increasing sequence (f_n) in $\mathscr{H}$ with the following property: if we denote by f_∞ the upper envelope of the sequence (f_n) then

$$\rho(f_\infty) \geqslant u \geqslant f_\infty.$$

Corollary. – *Let $\mathscr{H} \subset \mathscr{A}$ be a set directed for the relation $\leqslant$ and satisfying the following three conditions:*

i) $0 \leqslant f \leqslant 1$ *for each* $f \in \mathscr{H}$;
ii) $f \leqslant \rho(f)$ *for each* $f \in \mathscr{H}$;
iii) $f \in \mathscr{H}$, $n \in N^*$ *imply* $\inf(nf, 1) \in \mathscr{H}$.

Then the upper envelope u of $\mathscr{H}$ belongs to $\mathscr{A}$, $u \leqslant \rho(u)$, $\tilde{u} = \sup\{\tilde{f} \mid f \in \mathscr{H}\}$ and u is a characteristic function.

Proof: The relations $u \in \mathscr{A}$, $u \leqslant \rho(u)$ and $\tilde{u} = \sup\{\tilde{f} \mid f \in \mathscr{H}\}$ follow from Theorem 3. To see that u is a characteristic function suppose that $u(x) > 0$ for some $x \in X$. There is then $f \in \mathscr{H}$ such that $f(x) > 0$ and for n large enough we have

$$1 = \inf(nf(x), 1) \leqslant u(x),$$

whence $u(x) = 1$. Thus u can assume only the values 0 and 1 and the assertion is proved.

CHAPTER IV

The existence of a lifting

Throughout this chapter we assume that $N=\bar{N}$. *With the exception of sections 3 and 4 we suppose that* $(X,N,\mathscr{R})$ *is strictly localizable and that* $\mathscr{C}$ *is a fixed partition of* X *consisting of non-negligible integrable sets satisfying* $\sup\{\tilde{K}|K\in\mathscr{C}\}=\tilde{X}$.

In section 1 we give some preliminary results. The existence of a lifting of $M^{\infty}(X,N,\mathscr{R})$ (for strictly localizable $(X,N,\mathscr{R})$) is given in section 2. In section 5 we extend the notion of lifting to functions with values in a completely regular space.

1. Several results concerning the extension of a lifting

We begin with the following:

Proposition 1. – *Let* $\mathscr{A}$ *be an admissible subalgebra of* M^{∞} *and assume that* $\mathscr{A}\supset\{\varphi_K|K\in\mathscr{C}\}$ *and* $\mathscr{A}\neq M^{\infty}$. *There is then a measurable set* A *such that* $\varphi_A\notin\mathscr{A}$. *Further*

$$\mathscr{A}'=\{\varphi_A f+\varphi_{\complement A}g|f\in\mathscr{A},\ g\in\mathscr{A}\}$$

is an admissible subalgebra of M^{∞}, $\mathscr{A}'\supset\mathscr{A}$ *and* $\mathscr{A}'\neq\mathscr{A}$.

Proof: The existence of a measurable set A such that $\varphi_A\notin\mathscr{A}$ is obvious. It is clear now that

$$\mathscr{A}'=\{\varphi_A f+\varphi_{\complement A}g|f\in\mathscr{A},\ g\in\mathscr{A}\}$$

is a subalgebra of M^{∞}, that $\mathscr{A}'\neq\mathscr{A}$ ($\varphi_A=\varphi_A\cdot 1+\varphi_{\complement A}\cdot 0\in\mathscr{A}'$ and $\varphi_A\notin\mathscr{A}$) and that $\mathscr{A}'\supset\mathscr{A}$; in particular $\mathscr{A}'$ contains 1 and $\mathscr{N}^{\infty}$. We shall now show that $\mathscr{A}'$ satisfies the conditions α) and β) of proposition 3, section 1, chapter 2.

To verify condition α), let (v_n) be a sequence of elements of $\mathscr{A}'$ which is bounded by some constant $a\geqslant 0$ and which converges pointwise to some function v. Then, for each n, $v_n = \varphi_A f_n + \varphi_{CA} g_n$ for some functions f_n and g_n in $\mathscr{A}$. By Proposition 1, section 1, chapter 2, we may assume that $|f_n| \leqslant a$, $|g_n| \leqslant a$; then $\limsup f_n \in \mathscr{A}$, $\limsup g_n \in \mathscr{A}$. It follows that $v = \varphi_A \limsup f_n + \varphi_{CA} \limsup g_n \in \mathscr{A}'$ and hence $\mathscr{A}'$ verifies condition α).

To verify condition β), let $\mathscr{F}' = \{F | F \text{ integrable}, \varphi_F \in \mathscr{A}'\}$. Clearly $\mathscr{F}' \supset \{K \cap A | K \in \mathscr{C}\}$ and $\mathscr{F}' \supset \{K \cap CA | K \in \mathscr{C}\}$. Let now $u: X \to R$ be a bounded function satisfying $|u| \leqslant a$ for some constant $a \geqslant 0$ and such that $u\varphi_F \in \mathscr{A}'$ for every $F \in \mathscr{F}'$. In particular, for each $K \in \mathscr{C}$, $\varphi_A(u\varphi_K) \in \mathscr{A}'$; there exists then $f_K \in \mathscr{A}$ such that $\varphi_A(u\varphi_K) = \varphi_A f_K$ and we may assume without loss of generality that $|f_K| \leqslant a$ and f_K vanishes outside K (use the fact that $\varphi_K \in \mathscr{A}$ for each $K \in \mathscr{C}$). Define now the bounded function $f: X \to R$ by the relations:

$$f\varphi_K = f_K \quad \text{for each} \quad K \in \mathscr{C}.$$

By[1]) condition β) for the algebra $\mathscr{A}$, f belongs to $\mathscr{A}$. Moreover $\varphi_A u = \varphi_A f$. In a similar way we show the existence of a function $g \in \mathscr{A}$ such that $\varphi_{CA} u = \varphi_{CA} g$. It follows that

$$u = \varphi_A u + \varphi_{CA} u = \varphi_A f + \varphi_{CA} g \in \mathscr{A}'.$$

Thus $\mathscr{A}'$ satisfies also condition β). This completes the proof of the proposition.

Let $\mathscr{A}$ be an admissible subalgebra of M^∞ and ρ a lifting of $\mathscr{A}$. For each $E \in \mathscr{B}$ we define:

$$\mathscr{H}_E = \{f \in \mathscr{A} | 0 \leqslant f \leqslant 1, \ \rho(f) = f, \ \tilde{f} \leqslant \tilde{\varphi}_E\}.$$

Proposition 2. – *Let $\mathscr{A}$ be an admissible subalgebra of M^∞, ρ a lifting of $\mathscr{A}$ and $E \in \mathscr{B}$. Then the upper envelope of $\mathscr{H}_E$ is a characteristic function which we denote by φ_{E_∞}. We have:*

i) $\tilde{\varphi}_{E_\infty} \leqslant \tilde{\varphi}_E$.
ii) $\varphi_{E_\infty} = \rho(\varphi_{E_\infty}) \in \mathscr{H}_E$.
iii) $E_\infty \cap (CE)_\infty = \varnothing$.
iv) *If $h \in \mathscr{A}$ and $h\varphi_E \equiv 0$ then $\{x | \rho(h)(x) \neq 0\} \subset (CE)_\infty$.*

Proof: Let $E \in \mathscr{B}$. The set $\mathscr{H}_E$ satisfies the hypotheses of the corollary at the end of section 3, chapter 3. Hence the upper envelope u of $\mathscr{H}_E$ belongs to $\mathscr{A}$, $u \leqslant \rho(u)$ and u is a characteristic function: $u = \varphi_{E_\infty}$; moreover $\tilde{u} \leqslant \tilde{\varphi}_E$. This shows that $\rho(u) \in \mathscr{H}_E$ and hence $u = \rho(u)$. Thus i) and ii) are proved.

---

[1]) See Remark 5.1) at the end of section 1, chapter 2.

Statement iii) follows immediately from the relations:

$$\tilde{\varphi}_{E_\infty} \leqslant \tilde{\varphi}_E, \; \tilde{\varphi}_{(CE)_\infty} \leqslant \tilde{\varphi}_{CE}, \; \rho(\varphi_{E_\infty}) = \varphi_{E_\infty}, \; \rho(\varphi_{(CE)_\infty}) = \varphi_{(CE)_\infty}.$$

To verify iv) let $h \in \mathscr{A}$ be such that $h\varphi_E \equiv 0$. Then for a suitable constant $C > 0$ (for instance $C = 1/(\|h\|_\infty + 1)$)

$$C|\rho(h)| \in \mathscr{H}_{CE},$$

hence $C|\rho(h)| \leqslant \varphi_{(CE)_\infty}$. This completes the proof of Proposition 2.

Theorem 1. – *Let $\mathscr{A}$ be an admissible subalgebra of M^∞ and suppose that $\mathscr{A} \supset \{\varphi_K | K \in \mathscr{C}\}$ and $\mathscr{A} \neq M^\infty$. Assume that ρ is a lifting of $\mathscr{A}$. Let A be a measurable set such that $\varphi_A \notin \mathscr{A}$ and let*

$$\mathscr{A}' = \{\varphi_A f + \varphi_{CA} g | f \in \mathscr{A}, \; g \in \mathscr{A}\}.$$

Then $\mathscr{A}'$ is an admissible subalgebra of M^∞, $\mathscr{A}' \supset \mathscr{A}$, $\mathscr{A}' \neq \mathscr{A}$ and there exists a lifting ρ' of $\mathscr{A}'$ such that $\rho' | \mathscr{A} = \rho$.

Proof: The first part of the conclusion follows from Proposition 1. In order to prove the existence of ρ' we proceed as follows:

Let $p \in M^\infty$ be a function with the following three properties (we use here the notations of Proposition 2 above):

1) $p \equiv \varphi_A$;
2) $p^2 = p$ (i.e. p is a characteristic function);
3) $p | A_\infty = 1$ and $p | (CA)_\infty = 0$ (i.e. p "separates" the sets A_∞ and $(CA)_\infty$).

Such a function p exists (we may choose for instance[1]) $p = \varphi_{(A \cup A_\infty) \cap C((CA)_\infty)}$. Now for $u = \varphi_A f + \varphi_{CA} g \in \mathscr{A}'$ we define

$$\rho'(u) = p\rho(f) + (1-p)\rho(g).$$

Note that if $u = \varphi_A f + \varphi_{CA} g$ and $u_1 = \varphi_A f_1 + \varphi_{CA} g_1$ belong to $\mathscr{A}'$ and $u \equiv u_1$, then $\rho'(u) = \rho'(u_1)$. In fact we have:

$$\varphi_A(f - f_1) \equiv 0, \quad \varphi_{CA}(g - g_1) \equiv 0.$$

By statement iv) of Proposition 2,

$$\{x | \rho(f)(x) \neq \rho(f_1)(x)\} \subset (CA)_\infty, \quad \{x | \rho(g)(x) \neq \rho(g_1)(x)\} \subset A_\infty$$

and by properties 2) and 3) of p,

$$p\rho(f) + (1-p)\rho(g) = p\rho(f_1) + (1-p)\rho(g_1).$$

This shows simultaneously that ρ' is well-defined and that ρ' satisfies axiom (II) (the verification of the other axioms of a lifting is straightforward).

---

[1]) Clearly $p \leqslant \varphi_{A \cup A_\infty}$ and hence $\tilde{p} \leqslant \tilde{\varphi}_A$. Conversely $\tilde{\varphi}_{CA} \geqslant \tilde{\varphi}_{(CA)_\infty}$, whence $\tilde{\varphi}_A \leqslant \tilde{\varphi}_{C((CA)_\infty)}$ and therefore $\tilde{\tilde{\varphi}}_A \leqslant \tilde{p}$.

Let us show finally that $\rho'|\mathscr{A}=\rho$. We have $f\in\mathscr{A}\Rightarrow f=\varphi_A f+\varphi_{CA} f$ $\Rightarrow\rho'(f)=p\rho(f)+(1-p)\rho(f)=\rho(f)$.

Hence Theorem 1 is proved.

We next prove:

Theorem 2. – *Let $(\mathscr{A}_n)$ be an increasing sequence of admissible subalgebras of M^∞ such that $\mathscr{A}_1\supset\{\varphi_K|K\in\mathscr{C}\}$. For each n let ρ_n be a lifting of $\mathscr{A}_n$. Suppose that $\rho_n|\mathscr{A}_m=\rho_m$ if $m\leqslant n$. Let $\mathscr{A}_\infty$ be the admissible subalgebra generated by $\bigcup_n\mathscr{A}_n$. There is then a lifting ρ_∞ of $\mathscr{A}_\infty$ such that $\rho_\infty|\mathscr{A}_n=\rho_n$ for each n.*

Proof: For each n let P_n be the projection onto $\mathscr{A}_n$. Let P_∞ be the projection onto $\mathscr{A}_\infty$.

Let $f\in M^\infty$. For each n we have

$$\|\rho_n(P_n f)\|_\infty=N_\infty(P_n f)\leqslant N_\infty(f)$$

and thus the sequence $(\rho_n(P_n f))$ is bounded. Let now $\mathscr{U}$ be an *ultrafilter* on [1]) N^* finer than the Fréchet filter[2]). For $f\in\mathscr{A}_\infty$ and each $x\in X$ define

$$\gamma(f)(x)=\lim_{\mathscr{U}}\rho_n(P_n f)(x).$$

By Theorem 3, section 5, chapter 2, the sequence $((P_n f)(x))$ converges to $f(x)$ for almost every $x\in X$. Hence

$$\gamma(f)\in\mathscr{A}_\infty \quad\text{and}\quad \gamma(f)\equiv f.$$

It is immediate that $\gamma:\mathscr{A}_\infty\to\mathscr{A}_\infty$ is a linear lifting of $\mathscr{A}_\infty$ and that $\gamma|\mathscr{A}_n=\rho_n$ for each n. By Theorem 2, section 2, chapter 3, there is a lifting ρ_∞ of $\mathscr{A}_\infty$ with the property that $\rho_\infty|\mathscr{D}=\gamma|\mathscr{D}$ whenever $\mathscr{D}\subset\mathscr{A}_\infty$ is an admissible subalgebra such that $\gamma|\mathscr{D}$ is a lifting of $\mathscr{D}$. It follows that for each n, $\rho_\infty|\mathscr{A}_n=\gamma|\mathscr{A}_n=\rho_n$ and hence the theorem is proved.

2. The existence of a lifting of M^∞

We shall now state and prove:

Theorem 3. – *There exists a lifting of M^∞.*

Proof: Let $\mathscr{I}$ be the set of all pairs $(\mathscr{A},\rho_{\mathscr{A}})$ where $\mathscr{A}$ is an admissible subalgebra of M^∞ containing the set $\{\varphi_K|K\in\mathscr{C}\}$ and $\rho_{\mathscr{A}}:\mathscr{A}\to\mathscr{A}$ is a lifting of $\mathscr{A}$ such that $\rho_{\mathscr{A}}(\varphi_K)=\varphi_K$ for each $K\in\mathscr{C}$. Let us remark that

[1]) N^* is the set of all integers $1,2,\ldots$

[2]) The idea of using the ultrafilter $\mathscr{U}$ was suggested to us by the reading of [22].

the set $\mathscr{I}$ is non-void. In fact, let $\mathscr{A}_0$ be the set of all bounded functions $f: X \to R$ with the property that, for each $K \in \mathscr{C}$, f is equal to a constant almost everywhere on K. Clearly $\mathscr{A}_0$ is an admissible subalgebra of M^∞. Define now $\rho_{\mathscr{A}_0}$ as follows: Let $f \in \mathscr{A}_0$. For each $K \in \mathscr{C}$, f is equal to a constant a_K almost everywhere on K. We define

$$\rho_{\mathscr{A}_0}(f)(x) = a_K \quad \text{for} \quad x \in K.$$

It is easy to see that $\rho_{\mathscr{A}_0}: \mathscr{A}_0 \to \mathscr{A}_0$ is a lifting of $\mathscr{A}_0$ and that $\rho_{\mathscr{A}_0}(\varphi_K) = \varphi_K$ for each $K \in \mathscr{C}$. Hence $(\mathscr{A}_0, \rho_{\mathscr{A}_0}) \in \mathscr{I}$.

We now *order* the set $\mathscr{I}$ as follows:

We write $(\mathscr{F}, \rho_{\mathscr{F}}) \leqslant (\mathscr{G}, \rho_{\mathscr{G}})$ if $\mathscr{F} \subset \mathscr{G}$ and $\rho_{\mathscr{G}} | \mathscr{F} = \rho_{\mathscr{F}}$. This is clearly an order relation on $\mathscr{I}$.

To prove the theorem it is enough to show that the set $\mathscr{I}$ is *inductive* for this order relation. In fact, by Zorn's lemma, there is then a maximal element $(\mathscr{H}, \rho_{\mathscr{H}})$ in $\mathscr{I}$. By Theorem 1 we must have $\mathscr{H} = M^\infty$ and hence the existence of a lifting of M^∞ is established.

It remains then to show that every totally ordered part of $\mathscr{I}$ has a majorant in $\mathscr{I}$: Let $\Phi = (\mathscr{A}_j, \rho_j)_{j \in J}$ be a totally ordered family of elements of $\mathscr{I}$ (here $\rho_j = \rho_{\mathscr{A}_j}$ and $j' \leqslant j''$ if and only if $(\mathscr{A}_{j'}, \rho_{j'}) \leqslant (\mathscr{A}_{j''}, \rho_{j''})$). Let $\mathscr{A}_\infty$ be the admissible subalgebra of M^∞ spanned by $\bigcup_{j \in J} \mathscr{A}_j$. We have to distinguish two cases:

A) *There is no countable cofinal part in J.*

It is easy to see that in this case $\mathscr{A}_\infty$ coincides with the set of all bounded functions $f: X \to R$ such that $f \varphi_K \in \bigcup_{j \in J} \mathscr{A}_j$ for each $K \in \mathscr{C}$. We now define ρ_∞ as follows: Let $f \in \mathscr{A}_\infty$ and let $K \in \mathscr{C}$. Since $f \varphi_K \in \bigcup_{j \in J} \mathscr{A}_j$ there is $j \in J$ such that $f \varphi_K \in \mathscr{A}_j$. We set

$$\rho_\infty(f) | K = \rho_j(f \varphi_K) | K.$$

It is clear that $\rho_\infty(f)$ is well-defined for each $f \in \mathscr{A}_\infty$, that ρ_∞ maps $\mathscr{A}_\infty$ into $\mathscr{A}_\infty$, that it satisfies the axioms (I)–(VI) and that $\rho_\infty | \mathscr{A}_j = \rho_j$ for each $j \in J$; in particular $\rho_\infty(\varphi_K) = \varphi_K$ for each $K \in \mathscr{C}$. This shows that $(\mathscr{A}_\infty, \rho_\infty)$ is a majorant for Φ.

B) *There is a countable cofinal part I in J.*

We may suppose that I is the set of elements of an increasing sequence $(j(n))_{n \in N^*}$ (the case when J is finite is obvious). Note that in this case $\mathscr{A}_\infty$ is the admissible algebra spanned by

$$\bigcup_{n \in N^*} \mathscr{A}_{j(n)}.$$

By Theorem 2 there is a lifting ρ_∞ of $\mathscr{A}_\infty$ such that $\rho_\infty|\mathscr{A}_{j(n)}=\rho_{j(n)}$ for each $n\in N^*$. We deduce that $\rho_\infty|\mathscr{A}_j=\rho_j$ for each $j\in J$, which shows that $(\mathscr{A}_\infty,\rho_\infty)$ is a majorant of Φ and completes the proof of Theorem 3.

From Theorem 3 we immediately deduce the following:

Corollary. – *Let $\mathscr{A}$ be an admissible subalgebra of M^∞ containing the set $\{\varphi_K|K\in\mathscr{C}\}$. There exists a lifting of $\mathscr{A}$.*

A method or proof entirely similar to that of Theorem 3 yields also the following

Theorem 4. – *Let $\mathscr{A}$ be an admissible subalgebra of M^∞ containing the set $\{\varphi_K|K\in\mathscr{C}\}$. Every lifting ρ of $\mathscr{A}$ satisfying*[1]) *$\rho(\varphi_K)=\varphi_K$ for all $K\in\mathscr{C}$ can be extended to a lifting of M^∞.*

3. Equivalence of strict localizability with the existence of a lifting of M^∞

In the previous section we proved that if $(X,N,\mathscr{R})$ is strictly localizable, then there is a lifting of M^∞. We shall show now that the converse implication also holds.

Theorem 5. – *The following assertions concerning $(X,N,\mathscr{R})$ are equivalent:*

i) *$(X,N,\mathscr{R})$ is strictly localizable.*
ii) *There is a lifting of M^∞.*

Proof: The implication i) $\Rightarrow$ ii) was proved in section 2.

ii) $\Rightarrow$ i). Let ρ be a lifting of M^∞. Let $(\tilde{X}_i)_{i\in I}$ be a *maximal family* (for $\leqslant$) with the following properties:

a) X_i is integrable and non-negligible for each $i\in I$;
b) $\tilde{X}_i\cap\tilde{X}_j=\tilde{\varnothing}$ if $i\neq j$.

We deduce that:

a') $\rho(X_i)$ is integrable and non-negligible for each $i\in I$;
b') $\rho(X_i)\cap\rho(X_j)=\varnothing$ if $i\neq j$.

We next show that the set $X_0=\bigcup_{i\in I}\rho(X_i)$ is measurable and $X_0\equiv X$. It is obviously enough to show that X_0 is measurable (the fact that $X_0\equiv X$ follows from the maximality of the family $(\tilde{X}_i)_{i\in I}$). To prove

[1]) By a somewhat more involved argument it can be shown that *any* lifting of $\mathscr{A}$ can be extended to a lifting of M^∞.

that X_0 is measurable we must show that for each $B \in \mathscr{B}_0$, $X_0 \cap B$, or equivalently $X_0 \cap \rho(B)$, is measurable. Let $B \in \mathscr{B}_0$ and define

$$J = \{i \in I \mid \rho(B) \cap \rho(X_\iota) \neq \varnothing\}.$$

Since $\mu(B) < \infty$ and since $\mu(\rho(B) \cap \rho(X_\iota)) > 0$ for all $i \in J$, the set J is at most countable. It follows that

$$X_0 \cap \rho(B) = \bigcup_{\iota \in J} \rho(X_\iota) \cap \rho(B)$$

is measurable.

Let now i_0 be an element of I and let $\mathscr{C} = (K_\iota)_{\iota \in I}$ be the partition of X consisting of the following sets:

$$K_\iota = \begin{cases} \rho(X_\iota) & \text{if} \quad i \neq i_0, \\ \rho(X_{\iota_0}) \cup \complement X_0 & \text{if} \quad i = i_0. \end{cases}$$

It is clear that $\sup\{\tilde{K}_\iota \mid i \in I\} = \tilde{X}$. Thus $(X, N, \mathscr{R})$ is strictly localizable and the theorem is proved.

4. Non-existence of a linear lifting for the $\mathscr{L}^p$ spaces $(1 \leqslant p < \infty)$

We say that an element $\tilde{A} \in \mathscr{B}/\mathscr{N}_0$ is an *atom* if $\tilde{A} \neq \varnothing$ and if the relation $\tilde{B} \leqslant \tilde{A}$ implies $\tilde{B} = \tilde{A}$ or $\tilde{B} = \tilde{\varnothing}$.

We say that $\tilde{E} \in \mathscr{B}/\mathscr{N}_0$ is *diffuse* if $\tilde{E}$ does not contain any atom. We note that if $E \in \mathscr{B}_0$ and $\tilde{E}$ is diffuse then for every α satisfying the inequalities $0 \leqslant \alpha \leqslant \mu(E)$ there is a set $F \subset E$, $F \in \mathscr{B}$ such that $\mu(F) = \alpha$.

Definition 1. – *A mapping $\rho: \mathscr{L}^p \to \mathscr{L}^p$ is called a linear lifting of $\mathscr{L}^p$ if it satisfies the conditions:*

(1) $\rho(f) \equiv f$;
(2) $f \equiv g$ *implies* $\rho(f) = \rho(g)$;
(3) $f \geqslant 0$ *implies* $\rho(f) \geqslant 0$;
(4) $\rho(af + bg) = a\rho(f) + b\rho(g)$.

We shall show that in general for $1 \leqslant p < \infty$, there is no linear lifting of $\mathscr{L}^p$:

Theorem 6. – *Suppose that there is a non-negligible $E \in \mathscr{B}_0$ such that $\tilde{E}$ is diffuse. Then there is no linear lifting of $\mathscr{L}^p$.*

Proof: We reason by contradiction. Suppose that ρ is a linear lifting of $\mathscr{L}^p$. Clearly the equation

$$\rho(\tilde{f}) = \rho(f)$$

unambiguously defines $\rho(\tilde{f})$ for each $\tilde{f} \in L^p$. For each $x \in X$, the mapping $\tilde{f} \to \rho(\tilde{f})(x)$ of L^p into R is a *positive* linear form on L^p and hence continuous. Thus for each $x\in X$, there is $u_x\in(L^p)'$ such that

$$\rho(\tilde{f})(x)=\langle\tilde{f},u_x\rangle \quad\text{for}\quad \tilde{f}\in L^p.$$

Let $a=\mu(E)$. Since $\tilde{E}$ is diffuse, there exists for each integer $n\geqslant 1$ a partition $\{F_1^{(n)},\dots,F_n^{(n)}\}$ of E into measurable sets such that $\mu(F_1^{(n)}) =\dots=\mu(F_n^{(n)})=a/n$. Define

$$F^{(n)}=\bigcup_{j=1}^{n}\{x|\rho(\varphi_{F_j^{(n)}})(x)=1\}.$$

Clearly $F^{(n)}\equiv E$ for each $n\geqslant 1$. It follows that

$$\mu\left(E\cap\bigcap_{n=1}^{\infty}F^{(n)}\right)=\mu(E)>0.$$

Choose $x_0\in E\cap\bigcap_{n=1}^{\infty}F^{(n)}$. For each $n\geqslant 1$ there is j, $1\leqslant j\leqslant n$ such that

$$\rho(\varphi_{F_j^{(n)}})(x_0)=1,$$

whence $\langle\tilde{\varphi}_{F_j^{(n)}},u_{x_0}\rangle=1$. Since $u_{x_0}\in(L^p)'$ and $\lim_{n\to\infty}N_p(\varphi_{F_j^{(n)}})=0$ this gives the desired contradiction and hence completes the proof of the theorem.

Corollary. – *Under the hypothesis of Theorem 6 there is no mapping* $\rho:\mathscr{L}\to\mathscr{L}$ *satisfying the conditions:*

(1′) $\rho(f)\equiv f$;
(2′) $f\equiv g$ *implies* $\rho(f)=\rho(g)$;
(3′) $\rho(1)=1$;
(4′) $\rho(af+bg)=a\rho(f)+b\rho(g)$;
(5′) $\rho(fg)=\rho(f)\rho(g)$.

Proof: It is enough to note that if a mapping $\rho:\mathscr{L}\to\mathscr{L}$ satisfying (1′)–(5′) existed, then $\rho|\mathscr{L}^p$ would be a linear lifting of $\mathscr{L}^p$.

Remark. – In the case when $(X,N,\mathscr{R})$ is the "Lebesgue space" of the real line the result in the above corollary was proved by John von Neumann ([102]).

5. The extension of a lifting to functions with values in a completely regular space

Throughout this section ρ is a fixed lifting of $M^\infty(X,N,\mathscr{R})$.
Let E be a *completely regular space.*

Definition 2. – *A mapping $f:X\to E$ is said to be weakly measurable if for each $h\in C_R(E)$, the mapping $h\circ f:X\to R$ is measurable.*

Remarks. – 1) Let $\mathscr{E}$ be the tribe of all Baire subsets of E (= the smallest tribe with respect to which every $h \in C_R(E)$ is measurable). Let $f: X \to E$. The assertion that f is a *weakly measurable mapping* is equivalent with the assertion that f is a *measurable mapping of* $(X, \mathscr{B})$ into $(E, \mathscr{E})$ (i.e. $f^{-1}(A) \in \mathscr{B}$ for each $A \in \mathscr{E}$).

2) Suppose that X is a locally compact space, N is the essential upper integral corresponding to a positive Radon measure on X and $\mathscr{R} = \mathscr{K}(X)$. Suppose also that there exists a countable set in $C_R(E)$ separating the points of E. Then a mapping $f: X \to E$ with $f(X)$ relatively compact is weakly measurable if and only if it is measurable in the sense of Bourbaki.

For $f: X \to E$ and $g: X \to E$ weakly measurable mappings we write $f \equiv g(w)$, or simply $f \equiv g$ when there is no ambiguity, whenever

$$h \circ f \equiv h \circ g \quad \text{for each} \quad h \in C_R(E).$$

Example. – Let $f: X \to E$ and $g: X \to E$ be two weakly measurable mappings. If $f \equiv g$ it does not necessarily follow that $\{x \mid f(x) \neq g(x)\}$ is negligible. In fact, take for instance $X = [0,1]$, $(X, N, \mathscr{R})$ the corresponding Lebesgue space, $E = \prod_{t \in [0,1]} X_t$ with $X_t = [0,1]$ for each $t \in [0,1]$ and define $f = (f_t)_{t \in [0,1]}$ and $g = (g_t)_{t \in [0,1]}$ by $f_t = 0$ and

$$g_t(x) = \begin{cases} 1 & \text{if} \quad x = t, \\ 0 & \text{if} \quad x \neq t \end{cases}$$

for each $t \in [0,1]$. We wish to remark however that if $C_R(E)$ contains a countable set separating the points of E, then $f \equiv g \Leftrightarrow \{x \mid f(x) \neq g(x)\}$ is negligible.

We denote by $M_E^\infty(X, N, \mathscr{R})$ the set of all *weakly measurable* mappings $f: X \to E$ such that $f(X)$ is *relatively compact*.

When there is no ambiguity we shall write M_E^∞ instead of $M_E^\infty(X, N, \mathscr{R})$.

Definition 3. – *A mapping* $\rho': M_E^\infty \to M_E^\infty$ *is called a lifting of* M_E^∞ *associated with* ρ *if:*

(1) $\rho'(f) \equiv f$;

(2) $f \equiv g$ *implies* $\rho'(f) = \rho'(g)$;

(3) $\rho(h \circ f) = h \circ \rho'(f)$ *for all* $f \in M_E^\infty$, $h \in C_R(E)$.

Note that there exists at most *one* lifting associated with ρ[1]. In fact, suppose that ρ' and ρ'' are both liftings associated with ρ and let $f \in M_E^\infty$. By condition (3) we have $h \circ \rho'(f) = h \circ \rho''(f)$ for all $h \in C_R(E)$ and therefore $\rho'(f) = \rho''(f)$.

We shall prove in Theorem 7 that for every completely regular space E there exists a (unique) lifting of M_E^∞ associated with ρ.

---

[1]) It follows that if $E = R$ then the lifting associated with ρ is precisely ρ.

First however we shall make several remarks; in these remarks we suppose that E is a completely regular space and that ρ' is a lifting of M_E^∞ associated with ρ.

Remarks. – 1) *It is easy to see that* (1) *and* (2) *of Definition* 3 *are in fact consequences of* (3).

2) *Let* $f \in M_E^\infty$. *If* $K \subset E$ *is a compact set containing* $f(X)$, *then* K *contains also* $\rho'(f)(X)$.

In fact, for each $y \notin K$ let $h_y \in C_R(E)$ be such that $h_y(y)=1$, $h_y|K=0$. Then

$$h_y \circ f = 0 \Rightarrow \rho(h_y \circ f)=0 \Rightarrow h_y \circ \rho'(f)=0 \Rightarrow y \notin \rho'(f)(X).$$

3) *Let* $f \in M_E^\infty$ *be such that* $\rho'(f)=f$. *If* $C \subset E$ *is a closed set, then* $f^{-1}(C)$ *is measurable and* $\rho(f^{-1}(C)) \subset f^{-1}(C)$.

In fact let $\mathscr{H}=\{h \in C_R(E) | 0 \leqslant h \leqslant 1, h|C=1\}$. Then $\inf_{h \in \mathscr{H}} h \circ f = \varphi_{f^{-1}(C)}$, the set $\mathscr{H}$ is directed for the relation $\geqslant$ and $\rho(h \circ f)=h \circ f$ for each $h \in \mathscr{H}$. By Theorem 3, section 3, chapter 3, $\varphi_{f^{-1}(C)}: X \to R$ is measurable and $\rho(\varphi_{f^{-1}(C)}) \leqslant \varphi_{f^{-1}(C)}$.

4) *Let* $f_1 \in M_E^\infty, f_2 \in M_E^\infty, \ldots, f_n \in M_E^\infty$ *and let* $\varphi \in C_R(E^n)$. *Then*

$$\rho(\varphi \circ (f_1, \ldots, f_n)) = \varphi \circ (\rho'(f_1), \ldots, \rho'(f_n)).$$

In fact, let $K_i=\overline{f_i(X)}$ for each $1 \leqslant i \leqslant n$. By Remark 2) above it is enough to show that for every $h \in C_R\left(\prod_{1 \leqslant i \leqslant n} K_i\right)$

$$(*) \qquad \rho(h \circ (f_1, \ldots, f_n)) = h \circ (\rho'(f_1), \ldots, \rho'(f_n)).$$

Let A be the set of all $h \in C_R\left(\prod_{1 \leqslant i \leqslant n} K_i\right)$ of the form $h=h_1 \otimes h_2 \otimes \cdots \otimes h_n$ with $h_i \in C_R(K_i)$ $(1 \leqslant i \leqslant n)$. By the "continuity property" of ρ (formula (1), section 1, chapter 3) and the Stone-Weierstrass theorem it is enough to show that (*) holds for each $h \in A$. However for $h=h_1 \otimes h_2 \otimes \cdots \otimes h_n \in A$ ($h_i \in C_R(K_i)$ for $1 \leqslant i \leqslant n$) we obviously have

$$h \circ (f_1, \ldots, f_n) = (h_1 \circ f_1) \cdots (h_n \circ f_n) \Rightarrow \rho(h \circ (f_1, \ldots, f_n))$$
$$= \rho(h_1 \circ f_1) \cdots \rho(h_n \circ f_n) = (h_1 \circ \rho'(f_1)) \cdots (h_n \circ \rho'(f_n)) = h \circ (\rho'(f_1), \ldots, \rho'(f_n)).$$

Theorem 7. – *Let* ρ *be a lifting of* M^∞ *and* E *a completely regular space. Then there exists a unique lifting* ρ' *of* M_E^∞ *associated with* ρ.

Proof: We have already remarked (after Definition 3) that there is at most one lifting of M_E^∞ associated with ρ; it remains therefore to show that there exists one.

Let $f \in M_E^\infty$, let $K \subset E$ be a compact set containing $f(X)$ and let $C_R(K)$ be the algebra of continuous real-valued functions on K to R. Let $z' \in X$ be arbitrary and let $\chi_{z'}$ be the mapping $h \to \rho(h \circ f)(z')$ of $C_R(K)$ into R. Then $\chi_{z'}$ is a character of $C_R(K)$ and hence there exists a unique $z'' \in K$ satisfying

$$\rho(h \circ f)(z') = h(z'')$$

for all $h \in C_R(K)$. Define now $\rho'(f): X \to E$ by

$$\rho'(f)(z') = z'' \quad \text{for} \quad z' \in X.$$

Then $\rho'(f)$ is well defined, $\rho'(f)(X)$ is contained in K (hence it is relatively compact) and

$$h \circ \rho'(f) = \rho(h \circ f)$$

for all $h \in C_R(K)$ and hence for all $h \in C_R(E)$. Thus ρ' satisfies (3) of Definition 3. Since (1) and (2) are consequences of (3), (see Remark 1) above), the theorem is proved.

Thus we established the existence and uniqueness of the lifting of M_E^∞ associated with ρ. We shall usually employ the notation ρ_E for the lifting of M_E^∞ associated with ρ, or simply ρ when there is no ambiguity.

CHAPTER V

Topologies associated with lower densities and liftings

In this chapter we define and study the topologies associated with lower densities (see Definition 4, section 1, chapter 3) and particularly with liftings of an admissible subalgebra (see [54] and [57]).

Throughout this chapter we assume that $(X,N,\mathscr{R})$ *is localizable. With the exception of sections 4 and 6 we assume that* $\mathscr{A}$ *is an admissible subalgebra of* M^∞ *and* $\mathscr{F}=\{A|\varphi_A\in\mathscr{A}\}$.

1. The topology associated with a lower density

We begin with the following result:

Proposition 1. – *Let* $\theta\colon\mathscr{F}\to\mathscr{F}$ *be a lower density of* $\mathscr{F}$. *Define*

(1) $$\mathscr{T}_\theta=\{\theta(A)-N|A\in\mathscr{F},\ N\in\mathscr{N}_0\}.$$

Then:

1) $\mathscr{T}_\theta\subset\mathscr{F}$ *and* $\mathscr{T}_\theta$ *is a topology on* X. *If* $U\in\mathscr{T}_\theta$ *then* U *is non-void if and only if* U *is not negligible.*

2) *A set* $A\subset X$ *belongs to* $\mathscr{N}_0$ *if and only if* A *is closed and nowhere dense for* $\mathscr{T}_\theta$.

3) *A function* $f\colon X\to\bar{R}$ *is* $\mathscr{F}$*-measurable if and only if there is* $N\in\mathscr{N}_0$ *such that* f *is* $\mathscr{T}_\theta$*-continuous on* $\complement N$.

Proof: 1) Let $\theta(A)-N\in\mathscr{T}_\theta$ and $\theta(B)-M\in\mathscr{T}_\theta$ (with N and M in $\mathscr{N}_0$). By property (IV′) of a lower density

$$(\theta(A)-N)\cap(\theta(B)-M)=\theta(A\cap B)-(N\cup M)\in\mathscr{T}_\theta;$$

since θ is monotone increasing (consequence of (IV′)) we also have

$$(\theta(A)\cup\theta(B))-(N\cup M)\subset(\theta(A)-N)\cup(\theta(B)-M)\subset\theta(A)\cup\theta(B)\subset\theta(A\cup B).$$

Thus

$$(\theta(A)-N)\cup(\theta(B)-M)\in\mathscr{T}_\theta,$$

and $\mathscr{T}_\theta$ is closed under finite intersections and finite unions. To prove that $\mathscr{T}_\theta$ is closed under arbitrary unions, it is then enough to verity that for a *directed* (for $\subset$) family $(\theta(A_\iota)-N_\iota)_{i\in I}$ of sets belonging to $\mathscr{T}_\theta$, the set

$$\bigcup_{\iota\in I}\theta(A_\iota)-N_\iota$$

belongs also to $\mathscr{T}_\theta$. Let $\tilde{A}$ be the supremum of the family $(\tilde{A}_i)_{i\in I}$ in $\mathscr{B}/\mathscr{N}_0$; then $A\in\mathscr{F}$ (see condition α') of Proposition 3, section 1, chapter 2). We shall show that

$$(*)\qquad \bigcup_{i\in I}\theta(A_\iota)-N_\iota\in\mathscr{F} \quad\text{and}\quad \bigcup_{\iota\in I}\theta(A_\iota)-N_\iota\equiv\theta(A).$$

Since $\theta(A)\supset\theta(A_\iota)\supset\theta(A_\iota)-N_\iota$ for each $i\in I$, the relations (*) will obviously imply that $\bigcup_{\iota\in I}\theta(A_\iota)-N_i$ belongs to $\mathscr{T}_\theta$ and hence that $\mathscr{T}_\theta$ is a topology on X. In turn, to prove (*) it will be enough to show that for every $B\in\mathscr{F}_0=\{F|F\in\mathscr{B}_0,\ \varphi_F\in\mathscr{A}\}$,

$$(**)\qquad \bigcup_{\iota\in I}(\theta(A_\iota)-N_\iota)\cap B\in\mathscr{F} \quad\text{and}\quad \theta(A)\cap B\equiv\bigcup_{\iota\in I}(\theta(A_\iota)-N_\iota)\cap B$$

(use conditions β), β') of Proposition 3, section 1, chapter 2 and Remark 5.2) at the end of the same section). Let then $B\in\mathscr{F}_0$. Since $\theta(A)\cap B$ is in the class of the "supremum" of the family $(\tilde{A}_\iota\cap\tilde{B})_{\iota\in I}$, there is an increasing sequence $((\theta(A_{\iota_n})-N_{\iota_n})\cap B)_{1\leqslant n<\infty}$ such that

$$\theta(A)\cap B\equiv\bigcup_{n=1}^{\infty}(\theta(A_{\iota_n})-N_{\iota_n})\cap B$$

(see Theorem 4, section 2, chapter 1). On the other hand

$$\theta(A)\cap B\supset\bigcup_{\iota\in I}\theta(A_\iota)\cap B\supset\bigcup_{\iota\in I}(\theta(A_\iota)-N_\iota)\cap B\supset\bigcup_{n=1}^{\infty}(\theta(A_{\iota_n})-N_{\iota_n})\cap B$$

and comparing with the preceding formula we deduce (**). Thus $\mathscr{T}_\theta$ is a topology on X. It is clear from the definition of $\mathscr{T}_\theta$ that if $U\in\mathscr{T}_\theta$ and $U\neq\emptyset$ then U is not negligible.

2) Let $A\subset X$ be closed and nowhere dense for $\mathscr{T}_\theta$. Then $\complement A\in\mathscr{T}_\theta$ and hence $A\in\mathscr{F}$. Now $A\equiv\theta(A)\equiv\theta(A)\cap A$, $\theta(A)\cap A\in\mathscr{T}_\theta$ and $\theta(A)\cap A\subset A$. Since A is nowhere dense (and $\theta(A)\cap A$ is open) we deduce that $\theta(A)\cap A=\emptyset$, whence $A\in\mathscr{N}_0$.

Conversely let $A\in\mathscr{N}_0$. Then $\complement A=X-A=\theta(X)-A\in\mathscr{T}_\theta$; hence A is closed $\mathscr{T}_\theta$. By 1), A is also nowhere dense for $\mathscr{T}_\theta$.

3) Suppose first that $f:X\to\bar{R}$ is $\mathscr{T}_\theta$-continuous on $\complement N$, where $N\in\mathscr{N}_0$. For each $a\in R$ we have

$$\{x|f(x)>a\}=(\{x|f(x)>a\}\cap N)\cup(\{x|f(x)>a\}\cap\complement N).$$

Now $\{x|f(x)>a\}\cap N\in\mathcal{N}_0$ and $\{x|f(x)>a\}\cap \complement N\in\mathcal{T}_\theta$. We deduce that $\{x|f(x)>a\}\in\mathcal{F}$ and thus f is $\mathcal{F}$-measurable.

Conversely, suppose that $f: X\to\bar{R}$ is $\mathcal{F}$-measurable. Let us assume first that f is bounded. Since every bounded $\mathcal{F}$-measurable function is the uniform limit of a sequence of simple functions "based" on sets belonging to $\mathcal{F}$, it is enough to consider the case $f=\varphi_E$, where $E\in\mathcal{F}$. Since φ_E is constant on the open sets $\theta(E)\cap E$ and $\theta(\complement E)\cap\complement E$ and since

$$(\theta(E)\cap E)\cup(\theta(\complement E)\cap\complement E)\equiv X$$

the assertion is proved (for φ_E). The general case can be reduced to the case when f is bounded by composing with a homeomorphism of $\bar{R}$ onto $[-1,1]$ (for instance $t\to t/(1+|t|)$).

Note concerning the terminology. – If $\theta:\mathcal{F}\to\mathcal{F}$ is a lower density of $\mathcal{F}$, we shall call $\mathcal{T}_\theta$ the *density topology* associated with θ.

Remark. – Let $(X,N,\mathcal{R})$ be the "Lebesgue space" of the n-dimensional Euclidean space (i.e., $X=R^n$, N is the upper integral corresponding to Lebesgue measure μ on R^n and $\mathcal{R}=\mathcal{K}(R^n)$). The classical example of a lower density is that of the *lower Lebesgue* density. This is defined as follows: We say that $Q\subset R^n$ is an open *n-cube* if

$$Q=I_1\times\cdots\times I_n$$

where $I_j\subset R$, $j=1,2,\ldots,n$, are open intervals of equal length. For every $A\in\mathcal{B}$ we define

$$D_A(x)=\lim_{\mu(Q)\to 0}\frac{\mu(A\cap Q)}{\mu(Q)}\quad (Q\text{ open } n\text{-cube},\quad Q\ni x)$$

whenever this limit exists. If $A\in\mathcal{B}$, $x\in R^n$ and $D_A(x)$ exists and equals 1 we say that A has *density* 1 *at* x. Finally define for each $A\in\mathcal{B}$

$$\theta(A)=\{x\in R^n|D_A(x)\text{ exists and }=1\}.$$

It is not difficult to verify that θ is a lower density of $\mathcal{B}$; θ is usually called the lower Lebesgue ordinary metric density. The corresponding density topology $\mathcal{T}_\theta$ can then be described as follows: A set $U\subset R^n$ belongs to $\mathcal{T}_\theta$ if and only if U is measurable and has density 1 at each one of its points. Since a set $U\subset R^n$ open in the Euclidean topology is measurable and obviously has density 1 at each one of its points we deduce that the topology $\mathcal{T}_\theta$ *is finer than the Euclidean topology.* The density topology $\mathcal{T}_\theta$ was studied in detail in [42] and [43]. The real-valued continuous functions in the topology $\mathcal{T}_\theta$ are the approximately continuous functions and it was proved in [43] that the topology $\mathcal{T}_\theta$ is *completely regular.*

2. Construction of a lifting from a lower density using the density topology

The next result shows how to construct a lifting of $\mathscr{A}$ from a lower density θ of $\mathscr{F}$ by making use of the density topology $\mathscr{T}_\theta$. This result was suggested by a technique used by Dixmier in [28] (p. 177), to construct a lifting on the "Lebesgue space" of [0,1].

We make first several remarks concerning the algebra $\pi(\mathscr{A})$. It is clear that (with respect to the N_∞ norm) $\pi(\mathscr{A})$ is a commutative Banach algebra over R with unit element $\tilde{1}$ and that $N_\infty(\tilde{1})=\tilde{1}$. The space Z of all *characters*[1]) of $\pi(\mathscr{A})$ is non-void and *compact* for the topology induced by $\sigma((\pi(\mathscr{A}))', \pi(\mathscr{A}))$.

If for each $\tilde{f}\in\pi(\mathscr{A})$ we denote by $\Phi_{\tilde{f}}$ the mapping $\chi\to\chi(\tilde{f})$ of Z into R, then $\tilde{f}\to\Phi_{\tilde{f}}$ is an *isomorphism* of the Banach algebra $\pi(\mathscr{A})$ onto the Banach algebra $C_R(Z)$. Note also that for any *ideal* J of $\pi(\mathscr{A})$ with $J\neq\pi(\mathscr{A})$ there is a character of $\pi(\mathscr{A})$ which vanishes on J.

The above results can be obtained either by directly imbedding $\pi(\mathscr{A})$ in a convenient complex Banach algebra or by remarking that $\pi(\mathscr{A})$ satisfies the following two conditions ([3], [5])

(i) $N_\infty(\tilde{f}^2)=(N_\infty(\tilde{f}))^2$;

(ii) $N_\infty(\tilde{f}^2+\tilde{g}^2)\geqslant N_\infty(\tilde{f}^2)$ for all $\tilde{f}\in\pi(\mathscr{A})$ and $\tilde{g}\in\pi(\mathscr{A})$.

We may now state and prove the following:

Proposition 2. – *Let $\theta:\mathscr{F}\to\mathscr{F}$ be a lower density of $\mathscr{F}$. For each $x\in X$ let J_x be the set of all $\tilde{f}\in\pi(\mathscr{A})$ for which there is f in the class $\tilde{f}$ such that f is $\mathscr{T}_\theta$-continuous at x and $f(x)=0$. Then:*

1) J_x is a closed ideal of $\pi(\mathscr{A})$ and $J_x\neq\pi(\mathscr{A})$.

2) If for each $x\in X$ we let χ_x be a character of $\pi(\mathscr{A})$ vanishing on J_x, then the formula

$$\rho(f)(x)=\chi_x(\tilde{f}),\quad f\in\mathscr{A},\quad x\in X$$

defines a lifting of $\mathscr{A}$. Moreover if $f\in\mathscr{A}$ and $x\in X$ are such that f is $\mathscr{T}_\theta$-continuous at x, then $\rho(f)(x)=f(x)$.

Proof: The proof of 1) is immediate (note that $J_x\neq\pi(\mathscr{A})$ since $\tilde{1}\notin J_x$).

2) It is clear that the mapping ρ satisfies the axioms (II)–(VI) (section 1, chapter 3). It remains to show that ρ satisfies (I). We first show that if $f\in\mathscr{A}$ and $x\in X$ are such that f is $\mathscr{T}_\theta$-continuous at x, then $\rho(f)(x)=f(x)$; in fact, since $g=f-f(x)$ is $\mathscr{T}_\theta$-continuous at x

[1]) Z is usually called the Stone space corresponding to the Banach algebra $\pi(\mathscr{A})$.

and $g(x)=0$, we have $\tilde{g}\in J_x$. We deduce that (use the fact that $\rho(\lambda)=\lambda$ for each constant function λ)

$$0=\chi_x(\tilde{g})=\rho(g)(x)=\rho(f)(x)-f(x),$$

whence $\rho(f)(x)=f(x)$. Now by statement 3) of Proposition 1, given $f\in\mathscr{A}$, there is $N\in\mathscr{N}_0$ such that f is $\mathscr{C}_\theta$-continuous on $\complement N$; it follows that $\rho(f)|\complement N=f|\complement N$ and therefore $\rho(f)\equiv f$. This completes the proof of the proposition.

The following is an easy but very useful consequence of Proposition 2:

Corollary. – *Let $\theta\colon\mathscr{F}\to\mathscr{F}$ be a lower density of $\mathscr{F}$ and let θ' be the corresponding upper density of $\mathscr{F}$*[1]*. There is then a lifting ρ of $\mathscr{F}$ such that*

$$\theta(A)\subset\rho(A)\subset\theta'(A)$$

for all $A\in\mathscr{F}$. In particular, $\mathscr{C}_\theta\subset\mathscr{C}_\rho$.

Proof: Let ρ be the lifting given in Proposition 2. Let now $A\in\mathscr{F}$. Since $\varphi_{\theta(A)}$ is $\mathscr{C}_\theta$-continuous at every $x\in\theta(A)$, we have

$$\varphi_{\theta(A)}|\theta(A)=\rho(\varphi_A)|\theta(A),$$

or equivalently

$$\theta(A)\subset\rho(A).$$

The inclusion $\rho(A)\subset\theta'(A)$ follows by duality.

3. The topologies associated with a lifting

We assume in this section that $\rho\colon\mathscr{A}\to\mathscr{A}$ is a lifting of $\mathscr{A}$.

We recall that for $\tilde{f}\in\pi(\mathscr{A})$ the equation

$$\rho(\tilde{f})=\rho(f)$$

unambiguously defines $\rho(\tilde{f})$. For each $x\in X$ we define the character χ_x of $\pi(\mathscr{A})$ by

$$\chi_x(\tilde{f})=\rho(\tilde{f})(x),\qquad \tilde{f}\in\pi(\mathscr{A}).$$

With the lifting ρ we shall *associate two topologies* on X. The *topology* $\mathscr{T}_\rho$ is defined by taking for a *base* the set $\{\rho(A)|A\in\mathscr{F}\}$. Thus $U\in\mathscr{T}_\rho$ if and only if there is a family $(A_j)_{j\in J}$ of sets belonging to $\mathscr{F}$ such that

$$U=\bigcup_{j\in J}\rho(A_j).$$

---

[1]) We recall that θ' is defined by $\theta'(A)=\complement\theta(\complement A)$, for $A\in\mathscr{F}$.

The *second topology* is the topology $\mathscr{T}_\rho$ introduced in section 1:

$$\mathscr{T}_\rho=\{\rho(A)-N|A\in\mathscr{F},\ N\in\mathscr{N}_0\}.$$

Equivalently we may define $\mathscr{T}_\rho$ by

$$\mathscr{T}_\rho=\{U|U\in\mathscr{F},\ U\subset\rho(U)\}.$$

It is clear that $\mathscr{T}_\rho\subset\mathscr{T}_\rho$. The idea of defining the topologies $\mathscr{T}_\rho$ and $\mathscr{T}_\rho$ goes back to J. Oxtoby (see [71], p. 447).

For any topological space (Y,T) we denote by $C^b_{\mathbf{R}}(Y,T)$ the space of all bounded real-valued T-continuous functions on Y. We denote by $C_{\bar{\mathbf{R}}}(Y,T)$ the set of all T-continuous functions on Y to $\bar{R}$.

The basic properties of the topologies $\mathscr{T}_\rho$ and $\mathscr{T}_\rho$ can be summarized as follows:

Theorem 1. – 1) *For any $U\in\mathscr{T}_\rho$ (respectively $U\in\mathscr{T}_\rho$) the closure of U in the topology $\mathscr{T}_\rho$ (respectively $\mathscr{T}_\rho$) is $\rho(U)$. Hence the topologies $\mathscr{T}_\rho$ and $\mathscr{T}_\rho$ are extremally disconnected.*

2) *The topology $\mathscr{T}_\rho$ is Hausdorff if and only if the mapping $x\to\chi_x$ of X into $(\pi(\mathscr{A}))'$ is injective.*

3) *The topology $\mathscr{T}_\rho$ is uniformizable; whence $\mathscr{T}_\rho$ is completely regular if and only if the mapping $x\to\chi_x$ is injective.*

4) *We have $C^b_{\mathbf{R}}(X,\mathscr{T}_\rho)=C^b_{\mathbf{R}}(X,\mathscr{T}_\rho)=M$ where*

$$M=\{g\in\mathscr{A}|g=\rho(g)\}.$$

Hence $C_{\bar{\mathbf{R}}}(X,\mathscr{T}_\rho)=C_{\bar{\mathbf{R}}}(X,\mathscr{T}_\rho)$. Moreover $\mathscr{T}_\rho$ is the weakest topology on X for which every function in M is continuous.

5) *For every $\mathscr{F}$-measurable function $f: X\to\bar{R}$ there is a (unique) $\mathscr{T}_\rho$ (also $\mathscr{T}_\rho$)-continuous function $f^*: X\to\bar{R}$ such that $f^*\equiv f$.*

The proof of the theorem is straightforward (does not make use of the Stone space corresponding to $\pi(\mathscr{A})$). For completeness we give the details.

Proof: 1) We shall only prove the assertion for the topology $\mathscr{T}_\rho$. Let $U\in\mathscr{T}_\rho$. Since $\rho(U)(=\complement\rho(\complement U))$ is $\mathscr{T}_\rho$-closed and $U\subset\rho(U)$, we have $\bar{U}$ (= closure of U in the topology $\mathscr{T}_\rho$) $\subset\rho(U)$. Let now $z\in\rho(U)$. Let V be a neighborhood of z in the topology $\mathscr{T}_\rho$; we may assume that $V=\rho(A)$ with $A\in\mathscr{F}$. Then $z\in\rho(U)\cap\rho(A)=\rho(U\cap A)$ whence $\mu(U\cap V)=\mu(\rho(U\cap A))>0$. Thus $U\cap V\neq\emptyset$. Since the neighborhood V of z was arbitrary we deduce $z\in\bar{U}$. Hence $\bar{U}=\rho(U)$.

2) Suppose the topology $\mathscr{T}_\rho$ is Hausdorff and let $x\neq y$. There are then neighborhoods $V=\rho(A)\ni x$ and $W=\rho(B)\ni y$ with $V\cap W=\emptyset$. Then $\chi_x(\tilde{\varphi}_A)=1$, $\chi_x(\tilde{\varphi}_B)=0$, $\chi_y(\tilde{\varphi}_B)=1$, whence $\chi_x\neq\chi_y$. Conversely assume $x\to\chi_x$ is injective and let $s\neq t$. There is then a set $E\in\mathscr{F}$ such that $\chi_s(\tilde{\varphi}_E)\neq\chi_t(\tilde{\varphi}_E)$; we deduce that $\rho(E)$ and $\rho(\complement E)$ separate s and t.

3) We shall show that the topology $\mathcal{T}_\rho$ is uniformizable. Let $F \subset X$ be $\mathcal{T}_\rho$-closed and let $x_0 \in \complement F$. Since $\complement F = \bigcup_{j\in J} \rho(A_j)$ for some family $(A_j)_{j\in J}$ of sets belonging to $\mathcal{F}$, there is $j_0 \in J$ such that $x_0 \in \rho(A_{j_0})$. The function $f = \varphi_{\rho(A_{j_0})}$ is $\mathcal{T}_\rho$-continuous, $f: X \to [0,1]$ and f "separates" F and x_0. Thus $\mathcal{T}_\rho$ is uniformizable (see axiom (O_{IV}), [14], chapter 9, p. 17).

4) To see that $M \subset C_R^b(X, \mathcal{T}_\rho)$, it is enough to note that every simple function "based on representative sets" (i.e., $f = \Sigma a_j \varphi_{\rho(A_j)} = \rho(f)$) is $\mathcal{T}_\rho$ -continuous and that M is the closure of the set of all such simple functions in the supremum norm topology (use the "continuity property" of ρ). As $C_R^b(X, \mathcal{T}_\rho) \subset C_R^b(X, \mathcal{C}_\rho)$ obviously, it remains to prove $C_R^b(X, \mathcal{C}_\rho) \subset M$. Let $f \in C_R^b(X, \mathcal{C}_\rho)$; for every $a \in R$, $\{x | f(x) > a\} \in \mathcal{C}_\rho$ and hence f is $\mathcal{F}$-measurable, i.e. $f \in \mathcal{A}$. Now $\rho(f)$ belongs to M, whence by what was proved above, $\rho(f)$ is $\mathcal{C}_\rho$-continuous. Since $\rho(f) \equiv f$ the set $\{x | \rho(f)(x) \neq f(x)\}$ is open and negligible. Hence this set is void and thus $f = \rho(f) \in M$.

The case of continuous functions with values in $\bar{R}$ can be reduced to the case of bounded continuous functions by composing with a homeomorphism of $\bar{R}$ onto $[-1,1]$.

The assertion that $\mathcal{T}_\rho$ is the weakest topology on X for which every $g \in M$ is continuous follows from 3).

5) It is clearly enough to establish the assertion for f bounded and $\mathcal{F}$-measurable. However in this case the unique $\mathcal{T}_\rho$ (also $\mathcal{C}_\rho$)-continuous function equivalent with f is $f^* = \rho(f)$. This completes the proof of the theorem.

The following is an immediate consequence of Theorem 1:

Corollary. – *Suppose the topology $\mathcal{T}_\rho$ is Hausdorff. Consider the mapping $\chi: x \to \chi_x$ of X into Z, the Stone space corresponding to $\pi(\mathcal{A})$ and let $Z_0 = \{\chi_x | x \in X\}$. Then:*

1) The mapping $\chi: x \to \chi_x$ is a homeomorphism of $(X, \mathcal{T}_\rho)$ onto the subspace Z_0 of Z.

2) Z_0 is dense in Z.

3) Every $h \in C_R^b(Z_0)$ has a (unique) continuous extension to Z.[1])

Proof: 1) Let $(x_\alpha)_{\alpha\in I}$ be a family of elements of X (I a directed set) and $x \in X$. To show that $\chi: x \to \chi_x$ is a homeomorphism it is enough to note the following:

$$\lim_\alpha x_\alpha = x \quad \text{in the topology } \mathcal{T}_\rho$$

---

[1]) Z is the Stone-Cech compactification of $(X, \mathcal{T}_\rho)$.

if and only if

$$\lim_{\alpha} g(x_\alpha)=g(x) \quad \text{for each} \quad g\in C_R^b(X,\mathcal{T}_\rho).$$

On the other hand

$$\lim_{\alpha} \rho(f)(x_\alpha)=\rho(f)(x) \quad \text{for each} \quad f\in\mathscr{A}$$

if and only if

$$\lim_{\alpha} \chi_{x_\alpha}(\tilde{f})=\chi_x(\tilde{f}) \quad \text{for each} \quad \tilde{f}\in\pi(\mathscr{A})$$

that is, if and only if

$$\lim_{\alpha} \chi_{x_\alpha}=\chi_x \quad \text{in the topology of } Z_0.$$

2) It will be sufficient to show that if $u\in C_R(Z)$ vanishes on Z_0 then $u=0$. Let then $u\in C_R(Z)$ such that $u|Z_0=0$. There is $\tilde{f}\in\pi(\mathscr{A})$ such that (use the isomorphism of $\pi(\mathscr{A})$ onto $C_R(Z)$): $u(\chi)=\chi(\tilde{f})$ for all $\chi\in Z$. We have:

$$x\in X \Rightarrow \chi_x\in Z_0 \Rightarrow \chi_x(\tilde{f})=0 \Rightarrow \rho(\tilde{f})(x)=0;$$

thus $\tilde{f}=\tilde{0}$ and hence $u=0$.

3) Let $h\in C_R^b(Z_0)$. By 1) there is $g\in\mathscr{A}$ such that $h(\chi_x)=\rho(g)(x)$ for all $x\in X$. The function $\Phi_{\tilde{g}}\in C_R(Z)$ is the desired continuous extension of h to Z.

As will be seen in the next section, both situations $\mathcal{T}_\rho=\mathcal{C}_\rho$ and $\mathcal{T}_\rho\neq\mathcal{C}_\rho$ can occur.

4. An example

In order to illustrate more vividly the nature of the topologies $\mathcal{T}_\rho$ and $\mathcal{C}_\rho$ associated with a lifting we shall now specialize our setting: *We assume in this section that $(X,N,\mathscr{R})$ is the "Lebesgue space" of n-dimensional Euclidean space, we denote by T the Euclidean topology of R^n, by θ the lower Lebesgue density and by $\mathcal{C}_\theta$ the corresponding density topology* (see also the Remark in section 1). With these notations we have:

Theorem 2. – 1) *Let ρ be any lifting of M^∞ such that $\mathcal{C}_\theta\subset\mathcal{C}_\rho$. Then the topology $\mathcal{C}_\rho$ is completely regular and hence $\mathcal{T}_\rho=\mathcal{C}_\rho$.*

2) *There is a lifting ρ' of M^∞ such that $T\subset\mathcal{C}_{\rho'}$ and $\mathcal{T}_{\rho'}\neq\mathcal{C}_{\rho'}$.*

Proof: 1) We note first that there are liftings ρ of M^∞ satisfying $\mathcal{C}_\theta\subset\mathcal{C}_\rho$ (see the Corollary of Proposition 2 in section 2). Since $T\subset\mathcal{C}_\theta\subset\mathcal{C}_\rho$, for such a lifting ρ the topology $\mathcal{C}_\rho$ is Hausdorff. We shall show that $\mathcal{C}_\rho$ is uniformizable; by Theorem 1 this will imply that $\mathcal{T}_\rho=\mathcal{C}_\rho$.

Let $F \subset X$ be $\mathscr{T}_\rho$-closed and let $x_0 \in \complement F$. Then $F = \rho(A) \cup N$ with $A \in \mathscr{B}$, $N \in \mathscr{N}_0$ and $x_0 \notin \rho(A)$, $x_0 \notin N$. Since $\mathscr{T}_\theta$ is completely regular (see the Remark in section 1), since N is $\mathscr{T}_\theta$-closed and $x_0 \notin N$, there is $f: X \to [0,1]$ continuous $\mathscr{T}_\theta$ such that $f(x_0)=1$ and $f|N=0$. Let $g = f\varphi_{\rho(\complement A)}$. Then $g: X \to [0,1]$, g is continuous $\mathscr{T}_\rho$, $g(x_0)=1$ and $g|F=0$.

2) Let ρ be a lifting of M^∞ such that $T \subset \mathscr{T}_\rho$. Let $C \in \mathscr{N}_0$ be a set having an adherent point (in the Euclidean topology T) $a \notin C$. Let

$$I_C = \{\tilde{f} \in L^\infty \mid \rho(\tilde{f})|C = 0\}.$$

Let H_a be the set of all $\tilde{g} \in L^\infty$ for which there is g in the class $\tilde{g}$ such that g is continuous at a in the topology T and $g(a)=0$. Clearly I_C and H_a are ideals of L^∞. Let J be the ideal of L^∞ generated by I_C and H_a, that is $J = I_C + H_a$. The ideal J does not contain $\tilde{1}$. In fact, for any $\tilde{h} \in J$, $\tilde{h} = \tilde{f} + \tilde{g}$ with $\tilde{f} \in I_C$ and $\tilde{g} \in H_a$, there is an open Euclidean neighborhood U of a such that $y \in U$ implies $|g(y)| \leqslant \frac{1}{2}$ for all $y \in U$; we deduce (since $U \subset \rho(U)$ and $|\varphi_U g| \leqslant \frac{1}{2}$)

$$y \in U \cap C (\neq \varnothing) \Rightarrow |\rho(\tilde{h})(y)| = |\rho(\tilde{f})(y) + \rho(\tilde{g})(y)| = |\rho(\tilde{g})(y)| \leqslant \tfrac{1}{2},$$

whence

$$N_\infty(\tilde{1} - \tilde{h}) = \|\rho(\tilde{1}) - \rho(\tilde{h})\|_\infty \geqslant \tfrac{1}{2}.$$

Let now ω be a character of L^∞ vanishing on J and define ρ' by the formulas:

$$\rho'(\tilde{f})|\complement\{a\} = \rho(\tilde{f})|\complement\{a\}, \qquad \rho'(\tilde{f})(a) = \omega(\tilde{f}).$$

It is immediate that ρ' is a lifting and that $T \subset \mathscr{T}_{\rho'}$ (note that for $U \in T$, $\rho'(\varphi_U)(x) = 1$[1]) for every $x \in U$ and thus $U \subset \rho'(U)$). Further if $\tilde{f} \in L^\infty$ is such that $\rho'(\tilde{f})|C (= \rho(\tilde{f})|C) = 0$, then $\tilde{f} \in J$, whence $\rho'(\tilde{f})(a) = \omega(\tilde{f}) = 0$. Thus C and a cannot be separated by a $\mathscr{T}_{\rho'}$-continuous function and hence $\mathscr{T}_{\rho'}$ is not uniformizable. We conclude that $\mathscr{F}_{\rho'} \neq \mathscr{T}_{\rho'}$.

Remark. – With the same notations as above, we would like to point out that if ρ is any lifting of M^∞ such that $T \subset \mathscr{T}_\rho$, then $\mathscr{T}_\rho$ is *not normal* and a set $K \subset X$ is *$\mathscr{T}_\rho$-compact* if and only if it is finite (for the proof of this statement see [57], Theorem 3).

5. Liftings compatible with topologies

We shall next give the following:

Definition 1. – *Let $\rho: \mathscr{A} \to \mathscr{A}$ be a lifting of $\mathscr{A}$ and $T \subset \mathscr{F}$ a topology on X. We say that ρ is compatible with T if $T \subset \mathscr{T}_\rho$.*

---

[1]) Obviously $\omega(\tilde{\varphi}_{\complement U}) = 0$ if $U \ni a$.

Remarks. – 1) Every lifting ρ of $\mathscr{A}$ is compatible with $\mathscr{T}_\rho$ and $\mathscr{C}_\rho$.

2) The lifting ρ is compatible with the topology $T \subset \mathscr{F}$ if and only if $U \subset \rho(U)$ for every $U \in T$.

3) Let ρ_1, ρ_2 be two liftings of $\mathscr{A}$. If ρ_2 is compatible with $\mathscr{C}_{\rho_1}$, then $\rho_1 = \rho_2$.

Proposition 3. – *Let $\rho: \mathscr{A} \to \mathscr{A}$ be a lifting of $\mathscr{A}$ and $T \subset \mathscr{F}$ a topology on X. For each $A \in \mathscr{F}$ define*

$$D(A,T) = \{x \in X \mid V \in T, \quad V \ni x \Rightarrow V \cap A \text{ is not negligible}\}.$$

Then $D(A,T) \subset \bar{A}$ $(=$closure of A for the topology $T)$ for each $A \in \mathscr{F}$ and the following assertions are equivalent:

i) *ρ is compatible with T.*

ii) *$\rho(A) \subset D(A,T)$ for each $A \in \mathscr{F}$.*

Proof: The inclusion $D(A,T) \subset \bar{A}$ for each $A \in \mathscr{F}$ is obvious. Let us now prove that i)$\Leftrightarrow$ii).

i)$\Rightarrow$ii). Let $A \in \mathscr{F}$. Let $x \in \rho(A)$ and $V \in T$, $V \ni x$. Since $\rho(A) \in \mathscr{C}_\rho$, $V \in \mathscr{C}_\rho$ and $V \cap \rho(A) \neq \emptyset$ we deduce that $V \cap \rho(A)$, and hence also $V \cap A$, is not negligible. Thus $x \in D(A,T)$ and the inclusion $\rho(A) \subset D(A,T)$ is proved.

ii)$\Rightarrow$i). Let $F \subset X$ be T-closed. Since $D(F,T) \subset \bar{F} = F$ and $\rho(F) \subset D(F,T)$ we deduce $\rho(F) \subset F$. Let now $U \in T$. By what we just proved we have $\complement\rho(U) = \rho(\complement U) \subset \complement U$, whence $U \subset \rho(U)$. This completes the proof.

The notion of lifting compatible with a topology can be characterized by a *local property* as follows:

Proposition 4. – *Let $\rho: \mathscr{A} \to \mathscr{A}$ be a lifting of $\mathscr{A}$ and $T \subset \mathscr{F}$ a topology on X. The following assertions are equivalent:*

i) *ρ is compatible with T.*

ii) *For any $f \in \mathscr{A}$, $g \in \mathscr{A}$ and $x \in X$ with the property that f and g coincide almost everywhere on a neighborhood of x, there is a neighborhood of x on which $\rho(f)$ and $\rho(g)$ coincide.*

Proof: i)$\Rightarrow$ii). Let $f \in \mathscr{A}$, $g \in \mathscr{A}$ and $x \in X$ with the property that f and g coincide almost everywhere on a neighborhood V of x; we may assume $V \in T$. Then $\rho(\varphi_V f) = \rho(\varphi_V g)$ since $\varphi_V f \equiv \varphi_V g$. Since $V \subset \rho(V)$ we deduce:

$$z \in V \Rightarrow \rho(f)(z) = \rho(\varphi_V)(z)\,\rho(f)(z) = \rho(\varphi_V f)(z) = \rho(\varphi_V g)(z)$$
$$= \rho(\varphi_V)(z)\,\rho(g)(z) = \rho(g)(z),$$

and hence i)$\Rightarrow$ii) is proved.

ii)⇒i). Let $U \in T$. We must show that $U \subset \rho(U)$ or equivalently that $\varphi_U \leqslant \rho(\varphi_U)$. Let $f = \varphi_U$, $g = 1$. By assumption, for each $x \in U$ there is a neighborhood W_x of x, $W_x \subset U$ on which $\rho(f)$ and $\rho(g)$ coincide:

$$z \in W_x \Rightarrow \rho(\varphi_U)(z) = \rho(1)(z) = 1.$$

Since $U = \bigcup_{x \in U} W_x$, it follows that $\rho(\varphi_U)(z) = 1$ for all $z \in U$, whence $\rho(\varphi_U) \geqslant \varphi_U$. This completes the proof of the proposition.

In the theorem below, the notation $\mathscr{V}(x)$ (= the set of all neighborhoods of x) and the term fundamental system of neighborhoods of x *refer to the topology* T.

Using statement 4) of Theorem 1 (see section 3) we immediately obtain the following theorem:

Theorem 3. – *Let $\rho: \mathscr{A} \to \mathscr{A}$ be a lifting of $\mathscr{A}$ and $T \subset \mathscr{F}$ a topology on X which is uniformizable. Let $H \subset C^b_{\mathbf{R}}(X, T)$ be a set with the property that the weakest topology on X making every $f \in H$ continuous coincides with T. The following assertions are then equivalent:*

i) *ρ is compatible with T* (i.e., $U \subset \rho(U)$ for every $U \in T$).
ii) *$\rho(F) \subset F$ for every $F \subset X$ which is T-closed.*
iii) *$\rho(f) = f$ for every $f \in C^b_{\mathbf{R}}(X, T)$.*
iv) *$\rho(f) = f$ for every $f \in H$.*
v) *For each $x \in X$, $\mathscr{V}_\rho(x) = \{\rho(V) | V \in \mathscr{F}$ and $V \in \mathscr{V}(x)\}$ is a fundamental system of neighborhoods of x.*
vi) *$T \subset \mathscr{T}_\rho$.*

Proof: i) and ii) are clearly equivalent by duality.

i)⇒iii) is obvious since $C^b_{\mathbf{R}}(X, T) \subset C^b_{\mathbf{R}}(X, \mathscr{T}_\rho) = M$.

iii)⇒iv) obviously.

iv)⇒i). Statement iv) means that $H \subset C^b_{\mathbf{R}}(X, \mathscr{T}_\rho)$. The hypothesis on H then implies that $T \subset \mathscr{T}_\rho$.

Thus i), ii), iii), iv) are equivalent.

i) (≡ii))⇒v). Note that since T is uniformizable, T satisfies axiom (O_{III}) ([14], chapter 9, p. 18): For each $x \in X$ the set of all T-closed neighborhoods of x is a fundamental system of neighborhoods of x. Let now $x \in X$, $V \in \mathscr{V}(x)$ and let $W \in \mathscr{V}(x)$ such that $W = \overline{W} \subset V$. We have:

$$\mathring{W} \subset \rho(\mathring{W}) \subset \rho(W) \subset W \subset V$$

and hence v) holds.

v)⇒i). Let $U \in T$. For each $x \in U$ let $U_x = \rho(U_x) \in \mathscr{V}_\rho(x)$ such that $U_x \subset U$. We have:

$$U = \bigcup_{x \in U} U_x \Rightarrow \rho(U) \supset \bigcup_{x \in U} \rho(U_x) = \bigcup_{x \in U} U_x = U.$$

Thus the implication v)⇒i) is proved.

It remains to prove i) $\Leftrightarrow$ vi). Note that vi) $\Rightarrow$ i) is obvious.

i) ($\equiv$iii)) $\Rightarrow$ vi). Let $B = C_R^b(X,T)$ and note that $B \subset M = C_R^b(X,\mathcal{T}_\rho)$. Since T is the weakest topology on X for which every $g \in B$ is continuous we deduce $T \subset \mathcal{T}_\rho$. This completes the proof of Theorem 3.

Remark. – If $H \subset C_R^b(X,T)$ is such that the set $\{f^{-1}(U) | U \subset R$ open, $f \in H\}$ is a base for the topology T, then H satisfies the hypothesis in the statement of Theorem 3. This is the case for instance if (X,T) is locally compact and $H = \mathcal{K}_R(X,T)$ (= the set of all continuous functions on X having compact support). See also [57] and [71].

6. A remark concerning liftings for functions with values in a completely regular space

In this section we assume that ρ *is a fixed lifting of* $M^\infty(X,N,\mathcal{R})$ *and E is a completely regular space.*

Let $\rho_E: M_E^\infty \to M_E^\infty$ be the lifting associated with ρ and E; we recall that we proved in section 5, chapter 4 the existence and uniqueness of the lifting associated with ρ and E. Using Theorem 1 we obtain:

Theorem 4. – *Let* $M_E = \{g \in M_E^\infty | \rho_E(g) = g\}$. *Then:*

1) *The set of all* $\mathcal{T}_\rho$*-continuous mappings* $f: X \to E$ *such that* $f(X) \subset E$ *is relatively compact coincides with* M_E.

2) *The set of all* $\mathcal{C}_\rho$*-continuous mappings* $f: X \to E$ *such that* $f(X) \subset E$ *is relatively compact coincides with* M_E.

Proof: We note first (since E is completely regular) that given any topological space (Y,T), a mapping $u: Y \to E$ is T-continuous if and only if $h \circ u: Y \to R$ is T-continuous for every $h \in C_R(E)$.

For the purposes of the proof we denote by B_1 the set of all $\mathcal{T}_\rho$-continuous mappings $f: X \to E$ such that $f(X)$ is relatively compact and by B_2 the set of all $\mathcal{C}_\rho$-continuous mappings $f: X \to E$ such that $f(X)$ is relatively compact. It suffices to show that we have the inclusions: $M_E \subset B_1 \subset B_2 \subset M_E$.

Let $g \in M_E$. By condition (3) of the definition of a lifting associated with ρ and E (see Definition 2, section 5, chapter 4) we have

$$\rho(h \circ g) = h \circ g \quad \text{for all} \quad h \in C_R(E)$$

and hence $h \circ g: X \to R$ is $\mathcal{T}_\rho$-continuous for every $h \in C_R(E)$. We deduce that $g: X \to E$ is $\mathcal{T}_\rho$-continuous and thus $g \in B_1$.

Since the inclusion $B_1 \subset B_2$ is obvious it remains to prove the inclusion $B_2 \subset M_E$. Let $f \in B_2$ (note that f is obviously weakly measurable and therefore $f \in M_E^\infty$). Since $h \circ f : X \to R$ is bounded and $\mathcal{T}_\rho$-continuous for each $h \in C_R(E)$, we deduce

$$h \circ f = \rho(h \circ f) \quad \text{for all} \quad h \in C_R(E).$$

But we also have

$$h \circ \rho_E(f) = \rho(h \circ f) \quad \text{for all} \quad h \in C_R(E).$$

It follows that $h \circ f = h \circ \rho_E(f)$ for all $h \in C_R(E)$, whence $f = \rho_E(f) \in M_E$. This completes the proof of the theorem.

CHAPTER VI

Integrability and measurability for abstract valued functions

In this chapter we shall discuss the integrability and measurability of functions with values in a Banach space. The definitions and results of sections 1, 2 and 3 are somewhat similar to those in chapter 1. The definitions and results of sections 4, 5, 6 are based on the notion of lifting and are essential for the applications in the next chapter.

Throughout this chapter X is a set, N is a (regular) upper integral and $\mathcal{R} \subset \mathcal{F}^1(X,N)$ a set of functions satisfying (L_1), (L_2) and (L_3) (section 2, chapter 1).

We recall that if E is a Banach space then we denote by E' the *dual* of E (the space of continuous linear forms on E) and by E^* the *algebraic dual* of E (the space of all the linear forms on E). For $x \in E$ and $x' \in E^*$ we write $\langle x, x' \rangle = x'(x)$. If $f: X \to E$ we denote by $\langle f, x' \rangle$ the mapping $t \to \langle f(t), x' \rangle$ of X into R (for each $x' \in E^*$). If $g: X \to E^*$ we denote by $\langle x, g \rangle$ the mapping $t \to \langle x, g(t) \rangle$ of X into R (for each $x \in E$).

1. The spaces $\mathcal{L}_E^p$ and L_E^p $(1 \leqslant p < +\infty)$

Let E be a Banach space and let E^X be the set of all mappings $f: X \to E$. For each $f \in E^X$ we denote by $\|f\|$ the mapping $t \to \|f(t)\|$ of X into R.

A function $f \in E^X$ is called N-negligible if $N(\|f\|) = 0$; hence $f \in E^X$ is N-negligible if and only if $\|f\|$ is N-negligible.

When there is no ambiguity we shall say *negligible* instead of *N-negligible*.

Let $f \in E^X$. For each $1 \leqslant p < +\infty$ we define $N_p(f)$ by

$$N_p(f) = N(\|f\|^p)^{1/p} \quad (= N_p(\|f\|)).$$

We denote by $\mathcal{F}_E^p(X,N)$ the set of all $f \in E^X$ for which $N_p(f) < +\infty$. The restriction of N_p to $\mathcal{F}_E^p(X,N)$ will be denoted by the same symbol.

When there is no ambiguity we shall write $\mathcal{F}_E^p$ instead of $\mathcal{F}_E^p(X,N)$.

Theorem 1. – *For each* $1 \leqslant p < +\infty$ *the set* $\mathscr{F}_E^p$ *is a vector space;* N_p *is a seminorm on* $\mathscr{F}_E^p$ *and* $\mathscr{F}_E^p$ *is complete with respect to this seminorm.*

We shall always assume that $\mathscr{F}_E^p$ is endowed with the topology defined by the seminorm N_p; we shall call this topology the *topology of mean convergence of order* p (for $p=1$ we call this simply the *topology of mean convergence*) on $\mathscr{F}_E^p$. If $A \subset \mathscr{F}_E^p$ then we shall usually call topology of mean convergence of order p on A (or topology of mean convergence if $p=1$) the topology induced on A by the topology of $\mathscr{F}_E^p$.

Let now $\mathscr{R}_E$ be the set consisting of all functions on X to E of the form

$$f_1 a_1 + \cdots + f_n a_n$$

where $f_1 \in \mathscr{R}, \ldots, f_n \in \mathscr{R}$, $a_1 \in E, \ldots, a_n \in E$. Clearly:

(1) $\mathscr{R}_E$ is a vector space;

(2) $\mathscr{R}_E \subset \mathscr{F}_E^p$ for each $1 \leqslant p < +\infty$;

(3) $f \in \mathscr{R}_E \Rightarrow \|f\| \in \mathscr{L}^p$ for each $1 \leqslant p < +\infty$.

The assertion (1) is obvious. The assertion (2) follows from the relations

$$N_p\left(\sum_{j=1}^{n} f_j a_j\right) \leqslant N_p\left(\sum_{j=1}^{n} |f_j| \, \|a_j\|\right) \leqslant \sum_{j=1}^{n} \|a_j\| \, N_p(f_j) < +\infty.$$

To prove (3) we reason as follows. Let $f = \sum_{j=1}^{n} f_j a_j$ in $\mathscr{R}_E$. For each $t \in X$, $f(t)$ belongs to the finite dimensional space spanned by $\{a_1, \ldots, a_n\}$. Hence there exists a sequence (x'_n) of elements of E' such that

$$\|f(t)\| = \sup_n |\langle f(t), x'_n \rangle|$$

for all $t \in X$. Whence $\|f\|$ is measurable. Since $\|f\| \in \mathscr{F}^p$ we deduce $\|f\| \in \mathscr{L}^p$.

Definition 1. – *Let* $1 \leqslant p < +\infty$. *We define* $\mathscr{L}_E^p(X,N,\mathscr{R})$ *to be the closure of* $\mathscr{R}_E$ *in* $\mathscr{F}_E^p(X,N)$. *We define* $L_E^p(X,N,\mathscr{R})$ *to be the separated space associated with* $\mathscr{L}_E^p(X,N,\mathscr{R})$.

Since $\mathscr{R}_E$ is a vector space, $\mathscr{L}_E^p(X,N,\mathscr{R})$ and $L_E^p(X,N,\mathscr{R})$ are vector spaces. The *canonical (linear) mapping* of $\mathscr{L}_E^p(X,N,\mathscr{R})$ onto $L_E^p(X,N,\mathscr{R})$ will be denoted by $f \to \tilde{f}$. Note that if f and g belong to $\mathscr{L}_E^p(X,N,\mathscr{R})$ then $\tilde{g} = \tilde{f}$ if and only if the set $\{x \mid g(x) \neq f(x)\}$ is N-negligible.

Recall that the norm on $L_E^p(X,N,\mathscr{R})$ (which will also be denoted by N_p) is defined by

$$N_p(\tilde{f}) = N_p(f)$$

for all $\tilde{f} \in L_E^p(X,N,\mathscr{R})$.

When there is no ambiguity we shall write $\mathscr{L}_E^p$ instead of $\mathscr{L}_E^p(X,N,\mathscr{R})$ and L_E^p instead of $L_E^p(X,N,\mathscr{R})$.

When $E=R$ then we shall *usually* write $\mathscr{L}^p$ and $\mathscr{L}^p(X,N,\mathscr{R})$ instead of $\mathscr{L}_R^p$ and $\mathscr{L}_R^p(X,N,\mathscr{R})$; also we shall *usually* write L^p and $L^p(X,N,\mathscr{R})$ instead of L_R^p and $L_R^p(X,N,\mathscr{R})$.

From Definition 1 it follows immediately that if $f \in \mathscr{L}_R^p (1 \leqslant p < +\infty)$ and $a \in E$ then $fa \in \mathscr{L}_E^p$; since $\mathscr{L}_E^p$ is a vector space we deduce that if $f_1, \ldots, f_n$ belong to $\mathscr{L}_R^p$ and $a_1, \ldots, a_n$ belong to E, then

$$f_1 a_1 + \cdots + f_n a_n \in \mathscr{L}_E^p.$$

Let $A \subset X$, $f: A \to E$. We say that f is *defined almost everywhere* if $\complement A$ is negligible. If $g: X \to E$, we say that g is equivalent with f if $g(x) = f(x)$ almost everywhere.

A function f with values in E, defined almost everywhere, is said to be *p-integrable* $(1 \leqslant p < +\infty)$ *with respect to* $(N, \mathscr{R})$ if it is equivalent with a function $g \in \mathscr{L}_E^p(X, N, \mathscr{R})$. If $p = 1$ then instead of 1-integrable with respect to $(N, \mathscr{R})$ we shall usually say *integrable with respect to* $(N, \mathscr{R})$ or $(N, \mathscr{R})$*-integrable.*

If f is a function with values in E defined almost everywhere, we define $N_p(f) = N_p(g)$ $(1 \leqslant p < +\infty)$ if $g: X \to E$ and g is equivalent with f. If f is p-integrable with respect to $(N, \mathscr{R})$ and if $g \in \mathscr{L}_E^p(X, N, \mathscr{R})$ is equivalent with f, then we shall sometimes write $\tilde{f} = \tilde{g}$. We define mean convergence of order p in a natural way for functions that are p-integrable with respect to $(N, \mathscr{R})$.

When there is no ambiguity we shall say p-integrable instead of p-integrable with respect to $(N, \mathscr{R})$ and integrable instead of $(N, \mathscr{R})$-integrable.

Let $1 \leqslant p < +\infty$ and let $\mathscr{U} \subset \mathscr{L}_R^p$ be a set *dense* in $\mathscr{L}_R^p$. Let $\mathscr{U}_E$ be the set of all functions of the form

$$f_1 a_1 + \cdots + f_n a_n$$

where $f_1, \ldots, f_n$ belong to $\mathscr{U}$ and $a_1, \ldots, a_n$ belong to E. Then $\mathscr{U}_E$ is dense in $\mathscr{L}_E^p$.

Denote by $\mathscr{S}_E(X, N, \mathscr{R})$ the set of all functions of the form

$$\varphi_{B_1} a_1 + \cdots + \varphi_{B_n} a_n$$

where $B_1, \ldots, B_n$ are integrable sets and $a_1, \ldots, a_n$ belong to E. From the Corollary to Theorem 11, section 4, chapter 1, and the above remark it follows that $\mathscr{S}_E(X, N, \mathscr{R})$ *is dense in* $\mathscr{L}_E^p (1 \leqslant p < +\infty)$.

When there is no ambiguity we shall write $\mathscr{S}_E$ instead of $\mathscr{S}_E(X, N, \mathscr{R})$. If $E = R$ we shall *usually* write $\mathscr{S}$ and $\mathscr{S}(X, N, \mathscr{R})$ instead of $\mathscr{S}_R$ and $\mathscr{S}_R(X, N, \mathscr{R})$.

Several properties of the spaces $\mathscr{L}_E^p$ and L_E^p $(1 \leqslant p < +\infty)$ are given below:

Theorem 2. – *Let* $\mathscr{E} \subset \mathscr{L}_E^p$ *be a set dense in* $\mathscr{L}_E^p$. *Then for every* $f \in \mathscr{L}_E^p$ *there exists a sequence* (f_n) *of functions belonging to* $\mathscr{E}$ *and having the following properties:*

2.1) *The sequence* (f_n) *converges to* f *in mean of order* p;

2.2) *The sequence* $(f_n(x))$ *converges to* $f(x)$ *almost everywhere;*

2.3) *There is* $g: X \to \bar{R}_+$ *with* $N_p(g) < +\infty$ *such that* $\|f_n\| \leqslant g$ *for each* n.

In particular the result in Theorem 2 is valid for $\mathscr{E} = \mathscr{R}_E$ or $\mathscr{E} = \mathscr{S}_E$. Moreover if $\mathscr{E} = \mathscr{S}_E$ and if $\|f(x)\| \leqslant M$ for all $x \in X$, then we may suppose $\|f_n(x)\| \leqslant M$ for all $x \in X$ and all n.

Theorem 3. – *For each* $f \in \mathscr{L}_E^p$ $(1 \leqslant p < +\infty)$ *the function* $\|f\|$ *belongs to* $\mathscr{L}_R^p$. *Moreover the mapping* $f \to \|f\|$ *of* $\mathscr{L}_E^p$ *into* $\mathscr{L}_R^p$ *is uniformly continuous.*

A very useful result is the following:

Theorem 4 (Lebesgue). – *Let* (f_n) *be a sequence of functions belonging to* $\mathscr{L}_E^p$ *with the following properties:*

1) *The sequence* $(f_n(x))$ *converges almost everywhere to a limit* $f(x) \in E$;

2) *There is* $g: X \to \bar{R}_+$ *with* $N_p(g) < +\infty$ *such that for each* n, $\|f_n(x)\| \leqslant g(x)$ *almost everywhere.*

Then the function f *(defined almost everywhere) is p-integrable and the sequence* (f_n) *converges to* f *in mean of order* p.

We shall close this section with the following result:

Theorem 5. – *Let* $1 \leqslant p < +\infty$ *and let* $\varphi \in \mathscr{L}_R^p$ *and* $f \in \mathscr{L}_E^p$. *If one of the functions* φ, f *is bounded, then* $\varphi f \in \mathscr{L}_E^p$.

2. Measurable functions

The definition of measurable function on X to E is given as in the case of functions with values in $\bar{R}$.

Definition 2. – *A function* $f \in E^X$ *is called* $(N, \mathscr{R})$*-measurable if given any* $B \in \mathscr{B}_0(X, N, \mathscr{R})$ *there is a sequence* (g_n) *of functions belonging to* $\mathscr{R}_E$ *such that*

$$\lim_n g_n(x) = f(x)$$

N-almost everywhere on B. We denote by $\mathscr{L}_E(X, N, \mathscr{R})$ *the set of all* $(N, \mathscr{R})$*-measurable mappings belonging to* E^X.

From the definition it follows immediately that $\mathscr{L}_E(X, N, \mathscr{R})$ is a vector space and that $f \in \mathscr{L}_E(X, N, \mathscr{R})$ implies $\|f\| \in \mathscr{L}_R(X, N, \mathscr{R})$. We also deduce that if $f \in \mathscr{L}_R(X, N, \mathscr{R})$ and $g \in \mathscr{L}_E(X, N, \mathscr{R})$ then $fg \in \mathscr{L}_E(X, N, \mathscr{R})$.

Let $A \subset X$ be $(N, \mathscr{R})$-measurable and $f: A \to E$. We say that f is $(N, \mathscr{R})$-measurable if the mapping $f': X \to E$ defined by

$$f'(x) = \begin{cases} f(x) & \text{if } x \in A, \\ 0 & \text{if } x \notin A \end{cases}$$

is $(N, \mathscr{R})$-measurable.

When there is no ambiguity we say measurable instead of $(N, \mathscr{R})$-measurable and we write $\mathscr{L}_E$ instead of $\mathscr{L}_E(X, N, \mathscr{R})$.

When $E = R$, we shall *usually* write $\mathscr{L}$ or $\mathscr{L}(X, N, \mathscr{R})$ instead of $\mathscr{L}_R$ or $\mathscr{L}_R(X, N, \mathscr{R})$, respectively.

A very useful result is the following:

Theorem 6. – *Let* $1 \leqslant p < +\infty$ *and let* $f: X \to E$. *The following assertions are equivalent:*

i) *The function* f *is p-integrable;*

ii) *The function* f *is measurable and* $N_p(f) < +\infty$.

The proof of this theorem is similar to that of Theorem 9, section 4, chapter 1.

Theorem 7 (Egorov). – *Let* (f_n) *be a sequence of functions belonging to* $\mathscr{L}_E$ *and let* $f: X \to E$. *Then the following assertions are equivalent:*

i) *For every set* $B \in \mathscr{B}_0$ *the sequence* $(f_n(x))$ *converges to* $f(x)$ *almost everywhere on* B.

ii) *For every* $B \in \mathscr{B}_0$ *and* $\varepsilon > 0$ *there exists* $B_\varepsilon \in \mathscr{B}_0$, $B_\varepsilon \subset B$ *such that* $\mu(B - B_\varepsilon) \leqslant \varepsilon$ *and such that* $(f_n | B_\varepsilon)$ *converges uniformly to* $f | B_\varepsilon$.

Moreover, if i) $(\equiv$ ii)) *holds, then* f *is measurable.*

Theorem 7 can be proved as in the case of functions with values in R. Note also that the equivalence between i) and ii) can be deduced directly from Theorem 10, section 4, chapter 1.

Theorem 8. – *A function* $f: X \to E$ *is measurable if and only if given any integrable set* $K \subset X$ *there is a sequence* (s_n) *of functions belonging to* $\mathscr{S}_E$ *such that* $(s_n(x))$ *converges to* $f(x)$ *almost everywhere on* K.

Let now $E_1, E_2, \ldots, E_n$ be n $(\geqslant 1)$ Banach spaces and let

$$E = \prod_{j=1}^{n} E_j.$$

Let F be a Banach space and $h: E \to F$ be a continuous mapping. We have:

Theorem 9. – *If* $f_j \in \mathscr{L}_{E_j}$ *for each* $1 \leqslant j \leqslant n$, *the function* $h(f_1, f_2, \ldots, f_n)$ *belongs to* $\mathscr{L}_F$.

Note that if E is a Banach space, $f \in \mathscr{L}_E^p$ and $g \in \mathscr{L}_{E'}^q$ (where $1<p<+\infty$ and $1<q<+\infty$ are such that $1/p+1/q=1$) then, by Theorem 9, the mapping

$$\langle f,g\rangle : x \to \langle f(x), g(x)\rangle$$

is measurable and

$$N(\langle f,g\rangle) \leqslant N_1(\|f\|\,\|g\|) \leqslant N_p(f)\,N_q(g).$$

In particular $\langle f,g\rangle \in \mathscr{L}_R^1$.

We shall close this section with a theorem giving a criterion for measurability. We suppose in the rest of this section that F is a *Banach space*, $G \subset F'$ is a *vector space* and that *the couple* (F,G) *satisfies the condition*[1]*):*

(*) *For every* $a \in F$, $\|a\| = \sup_{z' \in G, \|z'\| \leqslant 1} |\langle a, z'\rangle|$.

Theorem 10. – *A function* $f: X \to F$ *belongs to* $\mathscr{L}_F$ *if and only if the following two conditions are satisfied:*

10.1) $\langle f, z'\rangle \in \mathscr{L}_R$ *for every* $z' \in G$;

10.2) *For every* $K \in \mathscr{B}_0$ *there exists a negligible set* $A \subset K$ *and a countable set* $H \subset F$ *such that*[2]*)* $f(t) \in \bar{H}$ *for all* $t \in K - A$.

Proof: It is easy to see that if $f \in \mathscr{L}_F$ then 10.1) and 10.2) are satisfied.

Conversely, suppose that 10.1) and 10.2) are satisfied. Let $K \in \mathscr{B}_0$ and let A and H be as in 10.2). We may and shall assume that $\bar{H}$ is a vector space. There is then a countable set G_0 contained in the unit ball of G (use condition (*)) such that

$$\|x\| = \sup_{z' \in G_0} |\langle x, z'\rangle|$$

for all $x \in \bar{H}$. Hence for every $a \in \bar{H}$ and $t \in K - A$ we have

$$\|f(t) - a\| = \sup_{z' \in G_0} |\langle f(t) - a, z'\rangle|.$$

Using 10.1) we deduce that the mapping $t \to \|f(t) - a\|$ of K into R is measurable.

Fix $n \in N^*$. For each $a \in H$ let

$$B_{n,a} = K \cap \{t \mid \|f(t) - a\| \leqslant 1/n\};$$

then $B_{n,a} \in \mathscr{B}_0$ and $\bigcup_{a \in H} B_{n,a} \equiv K$. Let now $(B'_{n,a})_{a \in H}$ be a family of disjoint sets belonging to $\mathscr{B}_0$ such that

$$B'_{n,a} \subset B_{n,a} \quad \text{and} \quad \bigcup_{a \in H} B'_{n,a} = \bigcup_{a \in H} B_{n,a}.$$

If

$$f_n = \sum_{a \in H} \varphi_{B'_{n,a}}\, a$$

---

[1]) Condition (*) is satisfied in each of the following cases: $F=E$, $G=E'$ or $F=E'$, $G=E$ (E a Banach space).

[2]) Here $\bar{H}$ is the closure of H in the norm topology.

then f_n is well defined, $f_n \in \mathscr{L}_F$ (clearly f_n is the pointwise limit of a sequence of functions in $\mathscr{S}_F$) and

$$\|f_n(t) - f(t)\| \leqslant 1/n$$

for all $t \in K - A$. Thus $(f_n(t))$ converges to $f(t)$ almost everywhere on K. Since $K \in \mathscr{B}_0$ was arbitrary we deduce that $f \in \mathscr{L}_F$.

Corollary. – *If F is of countable type then a function $f: X \to F$ belongs to $\mathscr{L}_F$ if and only if $\langle f, z' \rangle \in \mathscr{L}_R$ for every $z' \in G$.*

3. Further definitions and properties. The spaces $\mathscr{L}_E^\infty$ and L_E^∞

For $f: X \to E$ and $g: X \to E$ we write $f \equiv g$ whenever the set

$$\{t \mid f(t) \neq g(t)\}$$

is N-negligible.

Let now $\mathscr{N}_E(X,N)$ be the set of all N-negligible functions on X to E. Clearly $\mathscr{N}_E(X,N)$ is a *vector space*. Note also that for f and g in $\mathscr{L}_E(X,N,\mathscr{R})$ we have $f \equiv g$ if and only if $f - g$ belongs to $\mathscr{N}_E(X,N)$.

We shall write

$$L_E(X,N,\mathscr{R}) = \mathscr{L}_E(X,N,\mathscr{R})/\mathscr{N}_E(X,N)$$

and we shall denote by $\pi: f \to \tilde{f}$ the canonical mapping of $\mathscr{L}_E(X,N,\mathscr{R})$ onto the quotient space $L_E(X,N,\mathscr{R})$. If $h \in \mathscr{L}_R(X,N,\mathscr{R})$ and $f \in \mathscr{L}_E(X,N,\mathscr{R})$ we shall write $\pi(hf) = \tilde{h}\tilde{f}$.

Whenever there is no ambiguity, we shall write L_E instead of $L_E(X,N,\mathscr{R})$. If $E = R$ then we *usually* write L and $L(X,N,\mathscr{R})$, instead of L_R and $L_R(X,N,\mathscr{R})$.

It is clear that the canonical mapping of $\mathscr{L}_E^p$ onto L_E^p $(1 \leqslant p < +\infty)$ can be identified with the restriction of π to $\mathscr{L}_E^p$ and L_E^p can be identified with $\pi(\mathscr{L}_E^p)$.

Theorem 11. – *Let $\mathscr{C} \subset \mathscr{B}$ be such that* $\sup\{\tilde{K} \mid K \in \mathscr{C}\} = \tilde{X}$ *and let $f: X \to E$. Then:*

11.1) *The function f is measurable if and only if for each $K \in \mathscr{C}$, the function $\varphi_K f$ is measurable.*

11.2) *We have $f \equiv 0$ if and only if $\varphi_K f \equiv 0$ for each $K \in \mathscr{C}$.*

Note that 11.2) follows immediately from Theorem 15, section 5, chapter 1 (apply 15.2) to the function $\|f\|$).

A function $f: X \to E$ is called *locally N-negligible if* $\{x \mid f(x) \neq 0\}$ is locally N-negligible. Note that f is locally N-negligible if and only if $\|f\|$ is locally N-negligible.

Let $B_E^\infty(X)$ be the *vector space of all bounded* mappings of X into E. For each $f \in B_E^\infty(X)$ let

$$\|f\|_\infty = \sup_{t \in X} \|f(t)\|;$$

then $f \to \|f\|_\infty$ is a *norm* on $B_E^\infty(X)$ and $B_E^\infty(X)$ is a *Banach space* (when endowed with this norm).

If $E = R$, we usually write $B^\infty(X)$ instead of $B_R^\infty(X)$.

For $f \in B_E^\infty(X)$ we define

$$N_\infty(f) = N_\infty(\|f\|).$$

Then N_∞ is a *seminorm* on $B_E^\infty(X)$. Note that if $f \in B_E^\infty(X)$ then $N_\infty(f) = 0$ if and only if f *is locally N-negligible.*

We now denote by $\mathscr{L}_E^\infty(X,N,\mathscr{R})$ the *vector space* of all *bounded* functions $f \in \mathscr{L}_E(X,N,\mathscr{R})$ and by $\mathscr{N}_E^\infty(X,N,\mathscr{R})$ the vector space consisting of all *bounded locally N-negligible functions.* By $L_E^\infty(X,N,\mathscr{R})$ we denote the *quotient vector space*

$$\mathscr{L}_E^\infty(X,N,\mathscr{R})/\mathscr{N}_E^\infty(X,N,\mathscr{R}).$$

When there is no ambiguity we shall write

$\mathscr{L}_E^\infty$ instead of $\mathscr{L}_E^\infty(X,N,\mathscr{R})$,
L_E^∞ instead of $L_E^\infty(X,N,\mathscr{R})$,
$\mathscr{N}_E^\infty$ instead of $\mathscr{N}_E^\infty(X,N,\mathscr{R})$.

If $E = R$ we usually omit the letter E in these notations. For instance we shall usually write L^∞ instead of L_R^∞. *In chapter* 1, *section* 7, *we denoted* $M^\infty(X,N,\mathscr{R})$ *the space* $\mathscr{L}_R^\infty(X,N,\mathscr{R})$ (see also section 7 of the present chapter).

The restriction of N_∞ to $\mathscr{L}_E^\infty$ is clearly a seminorm on $\mathscr{L}_E^\infty$; we shall denote it by the same symbol. The corresponding *norm* on the quotient space L_E^∞ will be again denoted by N_∞; when endowed with this norm L_E^∞ is a *Banach space.*

Suppose now that

e) $$N = \bar{N}.$$

In this case if $f: X \to E$ is bounded we have

$$f \in \mathscr{N}_E^\infty \Leftrightarrow f \equiv 0.$$

Hence the canonical mapping of $\mathscr{L}_E^\infty$ onto L_E^∞ can be identified with the restriction of π to $\mathscr{L}_E^\infty$ and L_E^∞ can be identified with $\pi(\mathscr{L}_E^\infty)$ (we denote the canonical mapping of $\mathscr{L}_E^\infty$ onto L_E^∞ again by π).

4. The spaces $M_F^\infty[G]$ and $L_F^\infty[G]$

Let E be a vector space. We recall that a set $A \subset E$ is *convex* if the relations $x \in A$, $y \in A$ and $\lambda \in [0,1]$ imply $\lambda x + (1-\lambda) y \in A$; we recall also that $A \subset E$ is *equilibrated* (=*circled*) if the relations $x \in A$, $|\lambda| \leqslant 1$ imply $\lambda x \in A$. For every $B \subset E$ we denote by $c(B)$ the convex hull of B (= the intersection of all convex sets containing B) and by $e(B)$ the equilibrated hull of B (= the intersection of all equilibrated sets containing B). If E is endowed with a *locally convex topology* and $B \subset E$ then $\overline{c(e(B))}$ denotes of course the closed convex equilibrated hull of B.

Let now F be a *Banach space*, $G \subset F'$ *a vector space closed for the norm topology of* F' and suppose that *the couple* (F,G) *satisfies the following two conditions:*

($\star$) *For every* $a \in F$, $\|a\| = \sup_{z' \in G, \|z'\| \leqslant 1} |\langle a, z' \rangle|$;

($\overset{\star}{\star}$) *If* $A \subset F$ *is* $\sigma(F,G)$*-compact then* $\overline{c(e(A))}$ *is* $\sigma(F,G)$*-compact*[1]).

Remarks. – 1) Conditions ($\star$) and ($\overset{\star}{\star}$) are satisfied in each of the following cases: $F = E$, $G = E'$ or $F = E'$, $G = E$ (E a Banach space). 2) Condition ($\star$) clearly implies that G *separates* the points of F.

Below we shall consider F *endowed with the topology* $\sigma(F,G)$. We shall now introduce the following:

Definition 3. – *We denote by* $M_F^\infty[G; X, N, \mathcal{R}]$ *(or simply* $M_F^\infty[G]$ *when there is no ambiguity) the vector space of all mappings* $f: X \to F$ *having the following properties:*

i) $f(X) \subset F$ *is relatively compact;*
ii) $\langle f, z' \rangle \in M_R^\infty$ *for every* $z' \in G$.

It is clear that if $h \in M_R^\infty$ and $f \in M_F^\infty[G]$ then $hf \in M_F^\infty[G]$.

Note that by condition ($\star$) and the "uniform boundedness principle" we have for every $f \in M_F^\infty[G]$

$$\|f\|_\infty = \sup_{t \in X} \|f(t)\| < \infty;$$

thus $M_F^\infty[G] \subset B_F^\infty(X)$. It follows that $f \to \|f\|_\infty$ is a norm on $M_F^\infty[G]$, $f \to N_\infty(f)$ is a seminorm on $M_F^\infty[G]$ and that for each $f \in M_F^\infty[G]$ we have

$$N_\infty(f) \leqslant \|f\|_\infty.$$

For f and g in $M_F^\infty[G]$ we write

$$f \equiv g \ \ (w)$$

---

[1]) Throughout this section if $B \subset F$ then $\bar{B}$ denotes the closure of B in the topology $\sigma(F,G)$.

whenever $\langle f,z'\rangle \equiv \langle g,z'\rangle$ for every $z'\in G$. This defines an equivalence relation in $M_F^\infty[G]$ compatible with the vector space structure of $M_F^\infty[G]$.

Consider now the equivalence relation R_w in $M_F^\infty[G]$ defined as follows: If f and g are in $M_F^\infty[G]$ then $(f,g)\in R_w$ if and only if $\langle f,z'\rangle$ and $\langle g,z'\rangle$ coincide $\bar{N}$-almost everywhere, for every $z'\in G$. Again R_w is an equivalence relation in $M_F^\infty[G]$ compatible with the vector space structure of $M_F^\infty[G]$. We denote the corresponding quotient space by $L_F^\infty[G;X,N,\mathscr{R}]$ (or simply $L_F^\infty[G]$ when there is no ambiguity).

Note that if $N=\bar{N}$ then, for f and g in $M_F^\infty[G]$ we have

$$(f,g)\in R_w \Leftrightarrow f\equiv g \ (w).$$

In the rest of this section we suppose that $(X,N,\mathscr{R})$ *is strictly localizable and that* ρ *is a linear lifting of* $M_R^\infty(X,N,\mathscr{R})$.

Denote by $f\to\dot{f}$ the canonical mapping of $M_F^\infty[G]$ onto $L_F^\infty[G]$. For each $\dot{f}\in L_F^\infty[G]$ define

$$N_\infty(\dot{f})=\inf\{N_\infty(g)\,|\,g\in M_F^\infty[G],\ \dot{f}=\dot{g}\}.$$

It is clear that $N_\infty:\dot{f}\to N_\infty(\dot{f})$ is a *seminorm* on $L_F^\infty[G]$; it is shown below that N_∞ is in fact a *norm*.

We shall now introduce the following:

Definition 4. – *A mapping* $\rho':M_F^\infty[G]\to M_F^\infty[G]$ *is called a linear lifting of* $M_F^\infty[G]$ *associated with*[1]) ρ *if*

(1) $\rho'(f)\equiv f \ (w)$;

(2) $f\equiv g \ (w)$ *implies* $\rho'(f)=\rho'(g)$;

(3) $\rho(\langle f,z'\rangle)=\langle\rho'(f),z'\rangle$ *for all* $f\in M_F^\infty[G]$, $z'\in G$.

Note that there exists at most *one* linear lifting of $M_F^\infty[G]$ associated with ρ. Note also that if ρ' is a linear lifting of $M_F^\infty[G]$ associated with ρ, then $\rho':M_F^\infty[G]\to M_F^\infty[G]$ is a *linear mapping* and

(4) $$\|\rho'(f)\|_\infty\leqslant N_\infty(f).$$

The linearity of ρ' is obvious. To prove (4) we reason as follows: Let z' be an arbitrary element in G with $\|z'\|\leqslant 1$; then $\langle\rho'(f),z'\rangle=\rho(\langle f,z'\rangle)$, whence

$$|\langle\rho'(f),z'\rangle|\leqslant N_\infty(\langle f,z'\rangle)\leqslant N_\infty(f).$$

The inequality (4) now follows using (*).

We remark that in the case when ρ is a *lifting* of M_R^∞ (and $\rho':M_F^\infty[G]\to M_F^\infty[G]$ is associated with ρ) we have for every $h\in M_R^\infty$ and $f\in M_F^\infty[G]$

(5) $$\rho'(hf)=\rho(h)\,\rho'(f).$$

---

[1]) See also section 7 at the end of this chapter.

Proposition 1. – *There exists a unique linear lifting ρ' of $M_F^\infty[G]$ associated with ρ. Moreover for each $f \in M_F^\infty[G]$ we have*

$$N_\infty(\dot{f}) = N_\infty(\rho'(f)) = \|\rho'(f)\|_\infty .$$

Proof: Let $f \in M_F^\infty[G]$ and let $K = \overline{c(e(f(X)))}$. We may consider F as "canonically imbedded" in G^*. By condition ($\overset{*}{*}$), $K \subset F$ is $\sigma(F, G)$-compact, convex and equilibrated; hence there is a family $(z_i')_{i \in I}$ of elements in G such that

$$K = \bigcap_{i \in I} \{x \in G^* \mid |\langle x, z_i' \rangle| \leqslant 1\}.$$

For each $t \in X$, the mapping $z' \to \rho(\langle f, z' \rangle)(t)$ is a linear form on G; we shall denote it by $\rho'(f)(t)$. Hence

$$\rho(\langle f, z' \rangle) = \langle \rho'(f), z' \rangle, \quad \text{for} \quad z' \in G,$$

where $\rho'(f)$ is the mapping $t \to \rho'(f)(t)$ of X into G^*. From the relations

$$|\langle f, z_i' \rangle| \leqslant 1 \quad \text{for each} \quad i \in I$$

we deduce

$$|\langle \rho'(f), z_i' \rangle| = |\rho(\langle f, z_i' \rangle)| \leqslant 1 \quad \text{for each} \quad i \in I;$$

whence $\rho'(f)$ takes values in $K \subset F$. It follows that $\rho'(f) \in M_F^\infty[G]$ and that $\rho'(f) \equiv f$ (w).

Let now $g \equiv f$ (w). For every $z' \in G$ we have $\langle f, z' \rangle \equiv \langle g, z' \rangle$ and therefore

$$\langle \rho'(g), z' \rangle = \rho(\langle g, z' \rangle) = \rho(\langle f, z' \rangle) = \langle \rho'(f), z' \rangle .$$

Since $z' \in G$ was arbitrary we deduce that $\rho'(g) = \rho'(f)$. Hence ρ' is a linear lifting of $M_F^\infty[G]$ associated with ρ.

Let $f \in M_F^\infty[G]$. Since $\dot{\widehat{\rho'(f)}} = \dot{f}$, it is clear that

$$N_\infty(\dot{f}) \leqslant N_\infty(\rho'(f)).$$

On the other hand for any $g \in M_F^\infty[G]$ such that $\dot{g} = \dot{f}$ we have (use (4) above)

$$N_\infty(\rho'(f)) \leqslant \|\rho'(f)\|_\infty = \|\rho'(g)\|_\infty \leqslant N_\infty(g)$$

whence

$$N_\infty(\rho'(f)) \leqslant \|\rho'(f)\|_\infty \leqslant N_\infty(\dot{f}).$$

Therefore

$$N_\infty(\dot{f}) = N_\infty(\rho'(f)) = \|\rho'(f)\|_\infty$$

and the proposition is completely proved.

Note concerning the terminology. – We established above the existence and *uniqueness* of the lifting of $M_F^\infty[G]$ associated with ρ. To simplify the notation, from now on we shall write ρ instead of ρ'. Since it will be always clear from the context whether the functions we consider are real-valued or abstract-valued, this will not cause any ambiguity.

Remark. – On the basis of (2) we may define $\rho(\dot{f})$ for each $\dot{f} \in L_F^\infty[G]$ by

$$\rho(\dot{f}) = \rho(f).$$

This unambiguously defines the mapping $\rho: \dot{f} \to \rho(\dot{f})$ of $L_F^\infty[G]$ into $M_F^\infty[G]$. It is clear that this mapping is linear and that

$$N_\infty(\dot{f}) = \|\rho(\dot{f})\|_\infty \quad \text{for} \quad \dot{f} \in L_F^\infty[G]. \tag{6}$$

Corollary. – *The mapping $N_\infty: \dot{f} \to N_\infty(\dot{f})$ is a norm on $L_F^\infty[G]$.*

5. The case of the spaces $M_{E'}^\infty[E]$ and $L_{E'}^\infty[E]$

Let E be a *Banach space* and E' its *dual*. Note that in this case (with the notation of the previous section) the space $M_{E'}^\infty[E]$ is simply the vector space of all mappings $f: X \to E'$ such that $\langle z, f \rangle \in M_R^\infty$ for every $z \in E$.

It is clear that

$$\mathscr{L}_{E'}^\infty \subset M_{E'}^\infty[E]. \tag{α}$$

If E' is of *countable type*, it follows from the Corollary to Theorem 10 that every function $f \in M_{E'}^\infty[E]$ is *measurable;* hence

$$M_{E'}^\infty[E] = \mathscr{L}_{E'}^\infty.$$

It also follows that if f and g belong to $M_{E'}^\infty[E]$ then

$$(f,g) \in R_w \Leftrightarrow f - g \in \mathscr{N}_E^\infty$$

and hence

$$L_{E'}^\infty[E] = L_{E'}^\infty.$$

In the rest of this section we suppose that $(X, N, \mathscr{R})$ is strictly localizable and that ρ is a lifting (not only a linear lifting) of $M_R^\infty(X, N, \mathscr{R})$. Let us recall (see section 4) that we use the notation ρ also for the "abstract (linear) lifting" of $M_{E'}^\infty[E]$ corresponding to the (linear) lifting ρ of M_R^∞.

In the general case (E' not necessarily of countable type) we have:

Proposition 2. – *Let $f \in \mathscr{L}_{E'}^\infty$. Then $\rho(f) \in \mathscr{L}_{E'}^\infty$ and $\rho(f) \equiv f$.*

Proof: Let $\mathscr{C}$ be the set of all integrable parts $K \subset X$ having the following property: there exists a sequence (s_n) of functions belonging to $\mathscr{S}_{E'}$ which converges to f uniformly on K. We deduce from Theorem 7 that $\sup\{\tilde{K} | K \in \mathscr{C}\} = \tilde{X}$.

Let now $K \in \mathscr{C}$ and let (s_n) be a sequence of functions belonging to $\mathscr{S}_{E'}$ which converges to f uniformly on K. Then the sequence $(\varphi_K s_n)$ converges uniformly to $\varphi_K f$. By (4) and (5) in section 4 we can write

$$\|\varphi_{\rho(K)}\rho(f) - \varphi_{\rho(K)}\rho(s_n)\|_\infty = \|\rho(\varphi_K f) - \rho(\varphi_K s_n)\|_\infty$$
$$= \|\rho(\varphi_K f - \varphi_K s_n)\|_\infty \leqslant N_\infty(\varphi_K f - \varphi_K s_n)$$

for all n. Hence $(\varphi_{\rho(K)}\rho(s_n))$ converges uniformly to $\varphi_{\rho(K)}\rho(f)$.

Now $\rho(s_n) \in \mathscr{S}_{E'}$ and $\rho(s_n) \equiv s_n$ for all n; also $\varphi_{\rho(K)} \equiv \varphi_K$. It follows that

$$\varphi_K \rho(f) \equiv \varphi_K f$$

for all $K \in \mathscr{C}$. From 11.2) of Theorem 11 we deduce that $f \equiv \rho(f)$ and hence the proposition is proved.

Theorem 12. – *The space $L^\infty_{E'}$ can be canonically identified with a supspace of $L^\infty_{E'}[E]$.*

Proof: Let Φ be the mapping of $L^\infty_{E'}$ into $L^\infty_{E'}[E]$ defined by

$$\Phi(\tilde{f}) = \dot{f}.$$

It is easy to see that Φ is a well defined linear mapping of $L^\infty_{E'}$ into $L^\infty_{E'}[E]$. Let now $f \in \mathscr{L}^\infty_{E'}$. By Proposition 2

$$N_\infty(\tilde{f}) = N_\infty(\rho(f)).$$

By Proposition 1

$$N_\infty(\rho(f)) = N_\infty(\dot{f}).$$

Thus

$$N_\infty(\tilde{f}) = N_\infty(\dot{f})$$

and Theorem 12 is proved.

Remark. – We saw above that if E' is of *countable type* then

$$L^\infty_{E'}[E] = L^\infty_{E'}.$$

6. The spaces $\mathscr{L}^p_{E'}[E]$ and $L^p_{E'}[E]$ $(1 \leqslant p < +\infty)$

Let E be a *Banach space* and E' its *dual*. Let $1 \leqslant p < +\infty$; recall that for each $f \in (E')^X$ we define

$$N_p(f) = N(\|f\|^p)^{1/p}.$$

We shall now introduce the following

Definition 5. – *For each* $1 \leqslant p < +\infty$ *we denote by* $\mathscr{L}^p_{E'}[E; X, N, \mathscr{R}]$ *(or simply* $\mathscr{L}^p_{E'}[E]$ *when there is no ambiguity) the vector space of all mappings* $f: X \to E'$ *having the following properties:*

i) *the function* $\langle z, f \rangle$ *is measurable for each* $z \in E$;

ii) $N_p(f) < +\infty$.

It is clear that

$$(\alpha) \qquad \mathscr{L}^p_{E'} \subset \mathscr{L}^p_{E'}[E].$$

For f and g in $\mathscr{L}^p_{E'}[E]$ $(1 \leqslant p < +\infty)$ we write

$$f \equiv g \ (w)$$

whenever $\langle z, f \rangle \equiv \langle z, g \rangle$ for all $z \in E$. This defines an equivalence relation on $\mathscr{L}^p_{E'}[E]$ compatible with its vector space structure. *We denote the corresponding quotient space by* $L^p_{E'}[E; X, N, \mathscr{R}]$ *(or simply* $L^p_{E'}[E]$ *when there is no ambiguity) and the canonical mapping of* $\mathscr{L}^p_{E'}[E]$ *onto* $L^p_{E'}[E]$ *by* $f \to \hat{f}$.

If E' is of *countable* type, it follows from the Corollary to Theorem 10 that every function $f \in \mathscr{L}^p_{E'}[E]$ is *measurable;* hence

$$\mathscr{L}^p_{E'}[E] = \mathscr{L}^p_{E'}.$$

It also follows that if f and g belong to $\mathscr{L}^p_{E'}[E]$ then

$$f \equiv g \ (w) \Leftrightarrow f \equiv g,$$

that is, $\hat{f} = \tilde{f}$ for all $f \in \mathscr{L}^p_{E'}[E]$ and hence

$$L^p_{E'}[E] = L^p_{E'}.$$

Let $\mathscr{L}_{E'}[E; X, N, \mathscr{R}]$ (or simply $\mathscr{L}_{E'}[E]$ when there is no ambiguity) be the vector space of all $f \in (E')^X$ such that

$$\langle z, f \rangle \in \mathscr{L}_R$$

for all $z \in E$. Clearly $\mathscr{L}_{E'} \subset \mathscr{L}_{E'}[E]$.

Let $\mathscr{N}_{E'}[E]$ be the vector space of all $f \in \mathscr{L}_{E'}[E]$ such that

$$\langle z, f \rangle \equiv 0$$

for each $z \in E$. We shall write

$$L_{E'}[E] = \mathscr{L}_{E'}[E] / \mathscr{N}_{E'}[E]$$

and we shall denote by η: $f \to \hat{f}$ the canonical mapping of $\mathscr{L}_{E'}[E]$ onto $L_{E'}[E]$.

It is clear that the canonical mapping of $\mathscr{L}^p_{E'}[E]$ onto $L^p_{E'}[E]$ $(1 \leqslant p < +\infty)$ can be identified with $\eta | \mathscr{L}^p_{E'}[E]$ and $L^p_{E'}[E]$ with $\eta(\mathscr{L}^p_{E'}[E])$. If $(X, N, \mathscr{R})$ is strictly localizable similar remarks are valid for $M^\infty_{E'}[E]$ and $L^\infty_{E'}[E]$.

Note also that if E' is of *countable type* then

$$\mathscr{L}_{E'}[E] = \mathscr{L}_{E'};$$

it also follows that if f and g belong to $\mathscr{L}_{E'}[E]$ then $\eta(f)=\eta(g)$ if and only if $f \equiv g$; hence

$$L_{E'}[E] = L_{E'}.$$

In the rest of this section we suppose that $(X,N,\mathscr{R})$ is strictly localizable and that ρ is a lifting of $M^\infty_R(X,N,\mathscr{R})$.

For each $\dot{f} \in L^p_{E'}[E]$ define

$$N_p(\dot{f}) = \inf\{N_p(g) \mid g \in \mathscr{L}^p_{E'}[E],\ \dot{g} = \dot{f}\}.$$

It is clear that $N_p: \dot{f} \to N_p(\dot{f})$ is a *seminorm* on $L^p_{E'}[E]$; it is shown below that N_p is in fact a *norm*.

Denote by $\mathscr{Y}(X,N,\mathscr{R})$, or simply $\mathscr{Y}$ when there can be no ambiguity, the set of all *countable families* $\mathscr{C} = (A_j)_{j \in J}$ of sets belonging to $\mathscr{B}(X,N,\mathscr{R})$ having the following two properties:

(a) $A_{j'} \cap A_{j''} = \varnothing$ if $j' \neq j''$;
(b) $\sup\{\tilde{A}_j \mid j \in J\} = \tilde{X}$.

We note that the following properties are valid:

(1) *Let $1 \leqslant p < +\infty$ and $f: X \to R$ such that $N_p(f) < +\infty$. There is then $(A_j)_{j \in J} \in \mathscr{Y}$ such that $\varphi_{A_j} f$ is bounded, for each $j \in J$.*

Since N is *regular* there exists $g \in \mathscr{L}^1_R$, $g \geqslant 0$ such that

$$|f|^p \leqslant g.$$

It is enough then to take

$$A_j = \{t \mid j \leqslant g(t) < j+1\}$$

for $j \in N$.

(2) *If $\mathscr{C}' = (A_i)_{i \in I}$ and $\mathscr{C}'' = (B_j)_{j \in J}$ belong to $\mathscr{Y}$ then*

$$\mathscr{C}''' = (A_i \cap B_j)_{(i,j) \in I \times J}$$

belongs to $\mathscr{Y}$.

(3) *If $\mathscr{C}' = (A_j)_{j \in J}$ belongs to $\mathscr{Y}$ and $\mathscr{C}'' = (B_j)_{j \in J}$ is a family of disjoint sets belonging to $\mathscr{B}(X,N,\mathscr{R})$ such that*

$$A_j \equiv B_j$$

for each $j \in J$, then $\mathscr{C}'' \in \mathscr{Y}$.

(4) *If $\mathscr{C} = (A_j)_{j \in J}$ belongs to $\mathscr{Y}$ then $\rho(\mathscr{C}) = (\rho(A_j))_{j \in J}$ belongs to $\mathscr{Y}$.*

We shall now give the following:

Definition 6. – *For $f \in \mathscr{L}^p_{E'}[E]$ $(1 \leqslant p < +\infty)$ we write $\rho[f] = f$ whenever there is $\mathscr{C} = (K_j)_{j \in J}$ belonging to $\mathscr{Y}$ such that*

i) *$\varphi_{K_j}\langle z, f\rangle \in M^\infty_R$ for all $z \in E$ and $j \in J$;*
ii) *$\rho(\varphi_{K_j}\langle z, f\rangle) = \varphi_{\rho(K_j)}\langle z, f\rangle$ for all $z \in E$ and $j \in J$.*

Note that ii) implies that $\varphi_{\rho(K_j)}\langle z, f\rangle \in M^\infty_R$ for all $z \in E$ and $j \in J$.

Proposition 3. – 1) *For each* $f \in \mathscr{L}^p_{E'}[E]$ *there is* $g \in \mathscr{L}^p_{E'}[E]$ *such that* $f \equiv g$ (w) *and* $\rho[g] = g$*; moreover* $N_p(g) \leqslant N_p(f)$.

2) *If* g', g'' *are in* $\mathscr{L}^p_{E'}[E]$, $g' \equiv g''$ (w) *and* $\rho[g'] = g'$, $\rho[g''] = g''$, *then* g' *and* g'' *coincide almost everywhere.*

3) *If* $g \in \mathscr{L}^p_{E'}[E]$ *and* $\rho[g] = g$, *then* $N_p(g) = N_p(\dot{g})$. *Hence* $N_p : \dot{f} \to N_p(\dot{f})$ *is a norm on* $L^p_{E'}[E]$.

Proof: 1) Let $(K_j)_{j \in J} \in \mathscr{Y}$ be such that $\varphi_{K_j} \langle z, f \rangle \in M^\infty_R$ for every $z \in E$ and $j \in J$ (see (1)). Then $\varphi_{K_j} f \in M^\infty_{E'}[E]$ for every $j \in J$.

Let $g_j = \rho(\varphi_{K_j} f)$ for each $j \in J$ (see section 4 and 5) and let

$$g = \sum_{j \in J} g_j.$$

Clearly

$$\varphi_{\rho(K_j)} \langle z, g \rangle = \langle z, g_j \rangle = \rho(\varphi_{K_j} \langle z, f \rangle)$$

for all $z \in E$ and $j \in J$. We deduce that:

(5) $\varphi_{\rho(K_j)} \langle z, g \rangle \in M^\infty$ for all $z \in E$ and $j \in J$;

(6) $\rho(\varphi_{\rho(K_j)} \langle z, g \rangle) = \varphi_{\rho(K_j)} \langle z, g \rangle$ for all $z \in E$ and $j \in J$.

Since $(\rho(K_j))_{j \in J} \in \mathscr{Y}$ (see (4)) it follows from (5) and (6) that $g : X \to E'$ satisfies the conditions i) and ii) of Definition 6. Now it is clear that

$$\varphi_{\rho(K_j) \cap K_j} \langle z, g \rangle \equiv \varphi_{\rho(K_j) \cap K_j} \langle z, f \rangle$$

for all $z \in E$ and $j \in J$; since $(\rho(K_j) \cap K_j)_{j \in J} \in \mathscr{Y}$ we deduce that $\langle z, g \rangle \equiv \langle z, f \rangle$ for all $z \in E$.

To prove 1) completely it remains to show that $N_p(g) \leqslant N_p(f) (< +\infty)$.

For this let $H = \{j_1, \ldots, j_n\} \subset J$ and

$$f_H = (\varphi_{K_{j_1}} + \cdots + \varphi_{K_{j_n}}) f, \quad g_H = g_{j_1} + \cdots + g_{j_n}.$$

Note that

$$\langle z, g_H \rangle = \rho(\langle z, f_H \rangle)$$

for all $z \in E$. Let now $u \geqslant 0$ be an arbitrary *bounded integrable* function such that $\|f_H\|^p \leqslant u$ and let $u' = \rho(u)$. Since

$$|\langle z, f_H \rangle|^p \leqslant \|f_H\|^p \leqslant u$$

for all $z \in E$ with $\|z\| \leqslant 1$ we deduce

$$|\langle z, g_H \rangle|^p = |\rho(\langle z, f_H \rangle)|^p \leqslant u'$$

for all $z \in E$ with $\|z\| \leqslant 1$. Therefore $\|g_H\|^p \leqslant u'$ and hence

$$(N_p(g_H))^p = N(\|g_H\|^p) \leqslant N(u') = N(u).$$

Since u was arbitrary and N is regular we obtain

$$N_p(g_H) \leq N_p(f_H) \leq N_p(f).$$

Since $H \subset J$ was an arbitrary finite part we conclude

$$N_p(g) \leq N_p(f).$$

2) Let g', g'' be in $\mathscr{L}^p_{E'}[E]$ with $g' \equiv g''(w)$ and $\rho[g'] = g'$, $\rho[g''] = g''$. There are then $(K_i)_{i \in I} \in \mathscr{Y}$ and $(L_j)_{j \in J} \in \mathscr{Y}$ such that

$$\varphi_{K_i}\langle z, g'\rangle \in M^\infty, \quad \varphi_{L_j}\langle z, g''\rangle \in M^\infty \quad \text{for all } i \in I, \ j \in J \text{ and } z \in E;$$
$$\rho(\varphi_{K_i}\langle z, g'\rangle) = \varphi_{\rho(K_i)}\langle z, g'\rangle \quad \text{for all } z \in E \text{ and } i \in I;$$
$$\rho(\varphi_{L_j}\langle z, g''\rangle) = \varphi_{\rho(L_j)}\langle z, g''\rangle \quad \text{for all } z \in E \text{ and } j \in J.$$

Then $(\rho(K_i) \cap \rho(L_j))_{(i,j) \in I \times J} \in \mathscr{Y}$ and for each $(i,j) \in I \times J$ and $z \in E$ we have:

$$\varphi_{\rho(L_j)} \varphi_{\rho(K_i)}\langle z, g'\rangle = \rho(\varphi_{L_j \cap K_i}\langle z, g'\rangle) = \rho(\varphi_{\hat{K}_i \cap L_j}\langle z, g''\rangle)$$
$$= \varphi_{\rho(K_i)} \varphi_{\rho(L_j)}\langle z, g''\rangle.$$

We deduce that for each $(i,j) \in I \times J$

$$g' | \rho(K_i) \cap \rho(L_j) = g'' | \rho(K_i) \cap \rho(L_j),$$

whence g' and g'' coincide almost everywhere.

The assertion 3) follows from 1) and 2) and hence the proposition is proved.

Proposition 4. – *Let $f \in \mathscr{L}^p_{E'}$. Let $g' \in \mathscr{L}^p_{E'}[E]$ be such that $f \equiv g'$ (w) and $\rho[g'] = g'$. Then $f \equiv g'$ and hence $g' \in \mathscr{L}^p_{E'}$.*

Proof: As in the proof of 1), Proposition 3, let $(K_j)_{j \in J} \in \mathscr{Y}$ be such that $\varphi_{K_j}\langle z, f\rangle \in M^\infty$ for all $z \in E$ and $j \in J$. For each $j \in J$ let $g_j = \rho(\varphi_{K_j} f)$ and define

$$g = \sum_{j \in J} g_j.$$

Then $f \equiv g$ (w) and $\rho[g] = g$. Now for each $j \in J$, $\varphi_{K_j} f \in \mathscr{L}^\infty_{E'}$. By Proposition 2, $g_j \in \mathscr{L}^\infty_{E'}$ and $g_j \equiv \varphi_{K_j} f$. Since $\varphi_{\rho(K_j)} g = g_j$ and $\varphi_{\rho(K_j)} \equiv \varphi_{K_j}$ we deduce $\varphi_{K_j} g \equiv \varphi_{K_j} f$ for each $j \in J$. It follows that $g \equiv f$. The proof is completed by applying 2), Proposition 3.

Theorem 13. – *The space $L^p_{E'}$ $(1 \leq p < +\infty)$ can be canonically identified with a subspace of $L^p_{E'}[E]$.*

Proof: Let Φ be the mapping of $L^p_{E'}$ into $L^p_{E'}[E]$ defined by

$$\Phi(\tilde{f}) = \dot{f}.$$

It is easy to see that Φ is a well defined linear mapping of $L^p_{E'}$ into $L^p_{E'}[E]$. Let now $f \in \mathscr{L}^p_{E'}$ and let $g \in \mathscr{L}^p_{E'}[E]$ be such that $\rho[g] = g$ and $\dot{g} = \dot{f}$.

By Proposition 4, $f \equiv g$ and hence

$$N_p(\tilde{f}) = N_p(g).$$

By Proposition 3

$$N_p(g) = N_p(\dot{g}).$$

Thus

$$N_p(\tilde{f}) = N_p(\dot{f})$$

and Theorem 13 is proved.

Remark. – We saw above that if E' is of *countable type* then

$$L^p_{E'}[E] = L^p_{E'} \qquad (1 \leqslant p < +\infty).$$

7. A remark concerning the space $M_F^\infty[G]$

In this section we suppose that $(X, N, \mathscr{R})$ is strictly localizable.

Let us recall that for E a *completely regular* space we denoted by $M_E^\infty(X, N, \mathscr{R})$ (or simply M_E^∞) the set of all *weakly measurable* functions $f: X \to E$ such that $f(X) \subset E$ is *relatively compact* (see section 5, chapter 4).

Let now F be a *Banach space*, $G \subset F'$ a *vector space closed for the norm topology of* F' and suppose that *the couple* (F, G) *satisfies the conditions* $(\star)$ and $(\overset{\star}{\star})$ of section 4. We shall consider F endowed with the topology $\sigma(F, G)$; clearly $(F, \sigma(F, G))$ is *completely regular.*

For each $z' \in G$ denote by $h_{z'}$ the function $z \to \langle z, z' \rangle$ on F to R; note that $h_{z'} \in C_R((F, \sigma(F, G)))$. Let

$$\mathscr{H} = \{h_{z'} | z' \in G\}.$$

Then $\mathscr{H}$ separates the points of the completely regular space $(F, \sigma(F, G))$. We shall show that

$$M_F^\infty[G] = M^\infty_{(F, \sigma(F, G))}. \tag{1}$$

If $f \in M^\infty_{(F, \sigma(F, G))}$ then $f(X) \subset F$ is relatively $\sigma(F, G)$-compact and $h \circ f$ is measurable for each $h \in C_R((F, \sigma(F, G)))$. In particular

$$\langle f, z' \rangle = h_{z'} \circ f \in M_R^\infty$$

for all $z' \in G$, whence $f \in M_F^\infty[G]$.

Conversely, let $f \in M_F^\infty[G]$. Then $f(X) \subset F$ is relatively $\sigma(F, G)$-compact and if $h_1, \ldots, h_p$ belong to $\mathscr{H}$ then

$$(h_1 h_2 \ldots h_p) \circ f = (h_1 \circ f)(h_2 \circ f) \ldots (h_p \circ f)$$

is measurable. Since $\overline{f(X)}$ is $\sigma(F,G)$-compact we deduce (use the Stone-Weierstrass theorem) that $h \circ f$ is measurable for every $h \in C_R((F,\sigma(F,G)))$. Hence $f \in M^\infty_{(F,\sigma(F,G))}$ and (1) is proved.

From Theorem 7, section 5, chapter 4, it follows that if ρ is a lifting of M_R^∞ then there is a unique lifting of $M^\infty_{(F,\sigma(F,G))}$ associated with ρ, in the sense of Definition 3, section 5, chapter 4. Clearly if ρ' satisfies the properties (1), (2), (3) of this definition then $\rho': M_F^\infty[G] \to M_F^\infty[G]$ satisfies also the properties (1), (2), (3) of Definition 4, section 4 in this chapter.

Therefore the existence of a (linear) lifting of $M_F^\infty[G]$ associated with a lifting ρ of M_R^∞, in the sense of Definition 4 of this chapter, follows from the results of chapter 4. Note however that in chapter 4, the existence of ρ' was established under the hypothesis that ρ was a lifting, while in Proposition 1 of this chapter we assumed only that ρ was a linear lifting.

CHAPTER VII

Various applications

Throughout this chapter, with the exception of section 2, we suppose that $(X,N,\mathcal{R})$ *is strictly localizable.*

In section 1 we prove a general integral representation theorem without any separability assumptions. This theorem yields as corollaries the Dunford-Pettits theorem and the Dunford-Pettits-Phillips theorem (see also section 3). In section 2 we show that (for $(X,N,\mathcal{R})$ localizable) the existence of a linear lifting of M_R^∞ is equivalent with the validity of the Dunford-Pettits theorem in its most general form. Section 3 contains a result about weakly measurable and strongly measurable functions. In sections 4 and 5 we obtain the dual of L_E^p $(1 \leqslant p < +\infty)$ (without any separability hypotheses). Section 6 contains a proof of Strassen's integral representation theorem, again without any separability assumptions. Finally in section 7 we give an application to stochastic processes[1]).

Throughout this chapter we make constant use of the definitions, results and notations of chapter 6.

1. An integral representation theorem

In this section we suppose that ρ *is a linear lifting of* M_R^∞.

Let now F be a *Banach space*, $G \subset F'$ *a vector space closed for the norm topology of* F' and suppose that *the couple* (F,G) *satisfies the conditions* $(^\star)$ *and* $(^\star_\star)$ of section 4, chapter 6. With these notations we have:

Theorem 1. – 1) *Let* $U \in \mathcal{L}(L_R^1, F)$ *be such that* U *maps bounded sets of* L_R^1 *onto relatively* $\sigma(F,G)$*-compact sets of* F. *There is then a unique* $g_U \in M_F^\infty[G]$ *satisfying:*

---

[1]) For the integral representation of certain linear operators on certain L^p spaces $(1 \leqslant p < +\infty)$ see [27] and [67].

1.1) $\rho(g_U)=g_U$;

1.2) $\langle U\tilde{f},z'\rangle=\int f(t)\langle g_U(t),z'\rangle\,d\mu(t)$ *for all* $\tilde{f}\in L^1_R$ *and* $z'\in G$. *Moreover*[1])

1.3) $\overline{U(\{\tilde{f}\in L^1_R|N_1(\tilde{f})\leqslant 1\})}=\overline{c(e(g_U(X)))}$;

1.4) $N_\infty(\dot{g}_U)=\|g_U\|_\infty=\|U\|$.

2) *Conversely given any* $\dot{g}\in L^\infty_F[G]$ *there is a unique* $U\in\mathscr{L}(L^1_R,F)$ *mapping bounded sets of* L^1_R *onto relatively* $\sigma(F,G)$*-compact sets of* F *such that* $\dot{g}=\dot{g}_U$.

Proof: 1) Let $K=\overline{U(\{\tilde{f}\in L^1_R|N_1(\tilde{f})\leqslant 1\})}$. Note that K is $\sigma(F,G)$-compact, convex and equilibrated. We may consider F as "canonically imbedded" in G^*. There is then a family $(z'_\iota)_{\iota\in I}$ of elements in G such that

$$K=\bigcap_{\iota\in I}\{x\in G^*|\,|\langle x,z'_\iota\rangle|\leqslant 1\}.$$

Let $z'\in G$. The mapping $\tilde{f}\to\langle U\tilde{f},z'\rangle$ belongs to $(L^1_R)'=L^\infty_R$ and has norm $\leqslant\|U\|\,\|z'\|$. There is then $h_{z'}\in M^\infty_R$ such that $\|h_{z'}\|_\infty\leqslant\|U\|\,\|z'\|$ and

(1) $$\langle U\tilde{f},z'\rangle=\int f(t)h_{z'}(t)\,d\mu(t)$$

for all $\tilde{f}\in L^1_R$. It is clear that for any $\lambda\in R$, $x'\in G$ and $y'\in G$ we have:

$$h_{\lambda x'}\equiv\lambda h_{x'}\quad\text{and}\quad h_{x'+y'}\equiv h_{x'}+h_{y'}.$$

It follows that

(2) $$\|\rho(h_{z'})\|_\infty\leqslant\|U\|\,\|z'\|$$

for all $z'\in G$ and that

$$\rho(h_{\lambda x'})=\lambda\rho(h_{x'})\quad\text{and}\quad\rho(h_{x'+y'})=\rho(h_{x'})+\rho(h_{y'}).$$

Thus for each $t\in X$, the mapping $z'\to\rho(h_{z'})\,(t)$ is a linear form on G; we shall denote it by $g_U(t)$. Hence, for each $z'\in G$,

(3) $$\langle g_U,z'\rangle=\rho(h_{z'})\equiv h_{z'},$$

where g_U is the mapping $t\to g_U(t)$ of X into G^*.

Let now $\iota\in I$ be fixed. For each $\tilde{f}\in L^1_R$ with $N_1(\tilde{f})\leqslant 1$ we have by (1) and (3)

$$\left|\int f(t)\langle g_U(t),z'_\iota\rangle\,d\mu(t)\right|=|\langle U\tilde{f},z'_\iota\rangle|\leqslant 1.$$

Since $\tilde{f}\in L^1_R$ with $N_1(\tilde{f})\leqslant 1$ was arbitrary and since $\rho(\langle g_U,z'_\iota\rangle)=\langle g_U,z'_\iota\rangle$ we deduce

$$|\langle g_U(t),z'_\iota\rangle|\leqslant 1\quad\text{for all}\quad t\in X.$$

---

[1]) Below for a set $B\subset F$ we denote by $\bar{B}$ the closure of B in the topology $\sigma(F,G)$.

Since $i \in I$ is arbitrary it follows that g_U takes values in $K \subset F$. Thus

(4) $$\overline{c(e(g_U(X)))} \subset K = \overline{U\{\tilde{f} \in L^1_R | N_1(\tilde{f}) \leqslant 1\}},$$

$g_U \in M^\infty_F[G]$ and g_U satisfies 1.1) and 1.2). By (2) and (3) above and condition (*)

$$\|g_U\|_\infty \leqslant \|U\|.$$

On the other hand from 1.2) we have

$$|\langle U\tilde{f}, z'\rangle \leqslant N(|f| \, \|g_U\| \, \|z'\|) \leqslant \|g_U\|_\infty \|z'\| N_1(\tilde{f}).$$

Using again condition (*) we deduce that $\|U\| \leqslant \|g_U\|_\infty$; thus $\|g_U\|_\infty = \|U\|$ and 1.4) is proved (use also Proposition 1, section 4, chapter 6).

To establish the uniqueness assertion in 1) note that if g_1 and g_2 are two functions in $M^\infty_F[G]$ such that

$$\int f(t)\langle g_1(t), z'\rangle \, d\mu(t) = \int f(t)\langle g_2(t), z'\rangle \, d\mu(t)$$

for all $\tilde{f} \in L^1_R$ and $z' \in G$, then

$$g_1 \equiv g_2 \ (w).$$

2) Let $g \in M^\infty_F[G]$ and let $A = \overline{c(e(g(X)))}$. By condition ($\ast$), A is $\sigma(F, G)$-compact (also convex equilibrated). There is then a family $(x'_j)_{j \in J}$ of elements of G such that

$$A = \bigcap_{j \in J} \{x \in G^* | \, |\langle x, x'_j\rangle| \leqslant 1\}.$$

Note also that A is *bounded* (in norm). For $\tilde{f} \in L^1_R$ define $U\tilde{f}$ by

$$\langle U\tilde{f}, z'\rangle = \int f(t)\langle g(t), z'\rangle \, d\mu(t), \quad \text{for} \quad z' \in G.$$

Clearly $U\tilde{f}$ is a linear form on G. Moreover if $\tilde{f} \in L^1_R$ and $N_1(\tilde{f}) \leqslant 1$ we have for each $j \in J$

$$|\langle U\tilde{f}, x'_j\rangle| \leqslant \int |f(t)| \, |\langle g(t), x'_j\rangle| \, d\mu(t) \leqslant N_1(f) \leqslant 1,$$

whence $U\tilde{f}$ belongs to $A \subset F$. The inclusion

(5) $$U(\{\tilde{f} \in L^1_R | N_1(\tilde{f}) \leqslant 1\}) \subset A = \overline{c(e(g(X)))}$$

shows that $U \in \mathscr{L}(L^1_R, F)$ and that U maps bounded sets of L^1_R onto relatively $\sigma(F, G)$-compact sets of F. The equality $\dot{g} = \dot{g}_U$ follows from the uniqueness assertion in 1) and 1.3) from (4) and (5) above. This completes the proof of Theorem 1.

Let now E be a *Banach space*, E' the *dual of* E. From Theorem 1 we obtain the following important results:

Corollary 1 (Dunford-Pettis). – 1) *Let* $U \in \mathscr{L}(L_R^1, E')$. *There is then a unique* $g_U \in M_{E'}^\infty[E]$ *satisfying:*

i) $\rho(g_U) = g_U$;
ii) $\langle z, U\tilde{f}\rangle = \int f(t)\langle z, g_U(t)\rangle d\mu(t)$ *for all* $\tilde{f} \in L_R^1$ *and* $z \in E$.

Moreover

iii) $N_\infty(\dot{g}_U) = \|g_U\|_\infty = \|U\|$.

2) *Conversely given any* $\dot{g} \in L_{E'}^\infty[E]$ *there is a unique* $U \in \mathscr{L}(L_R^1, E')$ *such that* $\dot{g} = \dot{g}_U$.

3) *The mapping* $U \to \dot{g}_U$ *is an (isometric) isomorphism of the Banach space* $\mathscr{L}(L_R^1, E')$ *onto* $L_{E'}^\infty[E]$.

Remark. – Statement 3) of Corollary 1 shows that $L_{E'}^\infty[E]$ is a Banach space.

We recall that if E and F are Banach spaces, an operator $T \in \mathscr{L}(F, E)$ is called *weakly compact* if T maps bounded sets of F onto relatively $\sigma(E, E')$-compact sets of E. We write:

$$\mathscr{L}_w(F, E) = \{T \in \mathscr{L}(F, E) | T \text{ weakly compact}\}.$$

Corollary 2 (Dunford-Pettis-Phillips). – 1) *Let* $U \in \mathscr{L}_w(L_R^1, E)$. *There is then a unique* $g_U \in M_E^\infty[E']$ *satisfying:*

j) $\rho(g_U) = g_U$;
jj) $\langle U\tilde{f}, z'\rangle = \int f(t)\langle g_U(t), z'\rangle d\mu(t)$ *for all* $\tilde{f} \in L_R^1$ *and* $z' \in E'$.

Moreover

jjj) $\overline{U(\{\tilde{f} \in L_R^1 | N_1(\tilde{f}) \leqslant 1\})} = \overline{c(e(g_U(X)))}$;
jv) $N_\infty(\dot{g}_U) = \|g_U\|_\infty = \|U\|$.

2) *Conversely given any* $\dot{g} \in L_E^\infty[E']$ *there is a unique* $U \in \mathscr{L}_w(L_R^1, E)$ *such that* $\dot{g} = \dot{g}_U$.

3) *The mapping* $U \to \dot{g}_U$ *is an (isometric) isomorphism of the Banach space* $\mathscr{L}_w(L_R^1, E)$ *onto* $L_E^\infty[E']$.

Remarks. – 1) Since $\mathscr{L}_w(L_R^1, E)$ is a Banach space (see for instance [118]), statement 3) of Corollary 2 shows that $L_E^\infty[E']$ is a Banach space.

2) We shall see later (in section 3) that the function g_U in statement 1) of Corollary 2 is in fact *measurable*, i.e. belongs to $\mathscr{L}_E^\infty$.

2. The existence of a linear lifting of M_R^∞ is equivalent to the Dunford-Pettis theorem

We may now state the following theorem:

Theorem 2. – *Suppose that* $(X, N, \mathscr{R})$ *is localizable. Then the following assertions concerning* $(X, N, \mathscr{R})$ *are equivalent:*

i) *There exists a linear lifting of* M_R^∞.

ii) *Given any Banach space E and any* $U \in \mathscr{L}(L_R^1, E')$, *there is* $g \in M_{E'}^\infty[E]$ *such that* $\|g\|_\infty \leqslant \|U\|$ *and*

$$\langle z, U\tilde{f}\rangle = \int f(t) \langle z, g_U(t)\rangle \, d\mu(t)$$

for all $\tilde{f} \in L_R^1$ *and* $z \in E$.

Proof: The implication i)⇒ii) follows from Corollary 1 to Theorem 1. To prove the implication ii)⇒i) we reason as follows: Let $E = L_R^\infty$ and let U be the "canonical imbedding" of L_R^1 into $(L_R^1)'' = E'$. By ii) there is $g \in M_{E'}^\infty[E]$ such that $\|g(t)\| \leqslant 1 = \|U\|$ for each $t \in X$ and

$$\int f(t)\, u(t)\, d\mu(t) = \int f(t) \langle \tilde{u}, g(t)\rangle \, d\mu(t)$$

for all $\tilde{f} \in L_R^1$ and $\tilde{u} \in L_R^\infty$. We deduce that for each $u \in M_R^\infty$,

$$u \equiv \langle \tilde{u}, g\rangle .$$

Let $A = \{t | \langle \tilde{1}, g(t)\rangle = 1\}$. Then $\complement A$ is negligible. For each $t \in \complement A$ let χ_t be a character of L_R^∞. We define $\rho: M_R^\infty \to M_R^\infty$ as follows:

$$\rho(u)(t) = \begin{cases} \langle \tilde{u}, g(t)\rangle & \text{if} \quad t \in A, \\ \langle \tilde{u}, \chi_t\rangle & \text{if} \quad t \in \complement A \end{cases}$$

for each $u \in M_R^\infty$. It is clear that ρ satisfies the conditions (I), (II), (III) and (V) of a linear lifting (see section 1, chapter 3). It remains to show that ρ satisfies the positivity condition (IV). For this it is enough to remark that $\rho(1) = 1$, *that* $\|g(t)\| \leqslant 1$ for $t \in A$, $\|\chi_t\| \leqslant 1$ for $t \in \complement A$, and to use the fact that if $x' \in (L_R^\infty)'$ is such that $x'(\tilde{1}) = 1$ and $\|x'\| \leqslant 1$ then x' is positive. Hence Theorem 2 is proved.

3. Remarks concerning measurable functions and the spaces $M_E^\infty[E']$ and $L_E^\infty[E']$

In what follows, whenever F is a Banach space, the terms weak, weakly will always refer to the topology $\sigma(F, F')$.

For the discussion below we need the following:

Remark. – Let F be a Banach space and let $K \subset F$ be a *weakly compact set* and $H \subset F'$ be a *bounded set*. Consider K endowed with the weak topology. Then:

i) The set $\{z'|K| \; z' \in H\}$ is relatively weakly compact in $C_R(K)$;

ii) Given any sequence (z'_n) of elements of H, there is a subsequence (z'_{n_k}) and an element $z'_\infty \in F'$ such that

$$\lim_k z'_{n_k}(x) = z'_\infty(x) \quad \text{for all} \quad x \in K.$$

Proof: i) By a classical characterization of relatively weakly compact sets in $C_R(S)$ (S a compact space) (see [34], [45]) it suffices to show that if $(z_i')_{i\in I}$ (I a directed set) is a family of elements of H converging pointwise on K to some function h, then $h\in C_R(K)$. But this is immediate if we use the fact that H is relatively $\sigma(F',F)$-compact (Alaoglu's theorem).

ii) follows from i) and the Eberlein-Šmulian theorem (see [34]).

Let now $K\in\mathscr{B}_0$. Define

$$L_{R,K}^1=\{\tilde\varphi_K\tilde f\,|\,\tilde f\in L_R^1\}\quad\text{and}\quad L_{R,K}^\infty=\{\tilde\varphi_K\tilde f\,|\,\tilde f\in L_R^\infty\}.$$

Note that $L_{R,K}^\infty\subset L_{R,K}^1$ and that the topologies $\sigma(L_R^1,L_R^\infty)$ and $\sigma(L_{R,K}^1,L_{R,K}^\infty)$ induce on $L_{R,K}^1$ (and hence on each subset of $L_{R,K}^1$) the *same topology.*

For each $a>0$ define

$$B_K(a)=\{\tilde\varphi_K\tilde f\,|\,\tilde f\in L_R^\infty, N_\infty(\tilde f)\leqslant a\}.$$

Fix now $a>0$. Then $B_K(a)$ is $\sigma(L_R^1,L_R^\infty)$-*compact.* In fact, $B_K(a)$ is clearly $\sigma(L_R^\infty,L_R^1)$-compact or equivalently $\sigma(L_{R,K}^\infty,L_{R,K}^1)$-compact; hence $B_K(a)$ is also $\sigma(L_{R,K}^\infty,L_{R,K}^\infty)$-compact. Since $B_K(a)\subset L_{R,K}^\infty\subset L_{R,K}^1$ we deduce that $B_K(a)$ is $\sigma(L_{R,K}^1,L_{R,K}^\infty)$-compact and thus $\sigma(L_R^1,L_R^\infty)$-compact.

Let now E be a *Banach space.* We have:

Proposition 1. – *Let $U\in\mathscr{L}_w(L_R^1,E)$ and let $K\in\mathscr{B}_0$. Define*

$$S=\{U(\tilde\varphi_K\tilde f)\,|\,\tilde f\in L_R^1, N_1(\tilde f)\leqslant 1\}$$

and for each $a>0$

$$S(a)=\{U(\tilde\varphi_K\tilde f)\,|\,\tilde f\in L_R^\infty, N_\infty(\tilde f)\leqslant a\}.$$

Then:

1) *For each $a>0$ the set $S(a)$ is relatively compact in E.*
2) *There exists a countable set $S_0\subset S$ such that*[1] $\bar S_0=\bar S$.

Proof: 1) Let $a>0$. Note that

$$S(a)=U(B_K(a)).$$

Let now $(\tilde\varphi_K\tilde f_n)$ be a sequence of elements of $B_K(a)$. Since $B_K(a)$ is $\sigma(L_R^1,L_R^\infty)$-compact, there is an element $\tilde\varphi_K\tilde f$ in $B_K(a)$ and a subsequence of the sequence $(\tilde\varphi_K\tilde f_n)$ converging to $\tilde\varphi_K\tilde f$ for the topology $\sigma(L_R^1,L_R^\infty)$ (use the Eberlein-Šmulian theorem); to simplify the notation we shall assume that the sequence $(\tilde\varphi_K\tilde f_n)$ itself converges to $\tilde\varphi_K\tilde f$. We shall show that $(U(\tilde\varphi_K\tilde f_n))$ converges to $U(\tilde\varphi_K\tilde f)$ in E (for the norm topology), or equivalently, if we set $\tilde p_n=\tilde f-\tilde f_n$, that $(U(\tilde\varphi_K\tilde p_n))$ converges to 0.

---

[1]) Here the closure is taken in the norm topology of E.

Suppose this were false. There is then $\varepsilon>0$ such that $\|U(\tilde{\varphi}_K\tilde{p}_n)\|\geqslant\varepsilon$ for infinitely many n. Again for simplicity of notation we may assume that $\|U(\tilde{\varphi}_K\tilde{p}_n)\|\geqslant\varepsilon$ for all n. For each n choose $z'_n\in E'_1=\{z'\in E'\,|\,\|z'\|\leqslant 1\}$ such that

$$(1)\qquad |\langle U(\tilde{\varphi}_K\tilde{p}_n), z'_n\rangle|\geqslant\varepsilon/2.$$

Let $g=g_U$ be the function in $M_E^\infty[E']$ corresponding to U as in Corollary 2. By the Remark at the beginning of this section we may assume that there is $z'\in E'$ such that

$$(2)\qquad \lim_n\langle g(t),z'_n\rangle=\langle g(t),z'\rangle \quad \text{for all} \quad t\in X.$$

We have

$$|\langle U(\tilde{\varphi}_K\tilde{p}_n),z'_n\rangle|\leqslant|\int\varphi_K p_n\langle g,z'_n\rangle\,d\mu-\int\varphi_K p_n\langle g,z'\rangle\,d\mu|+|\int\varphi_K p_n\langle g,z'\rangle\,d\mu|.$$

The second term on the right tends to 0 since $(\tilde{\varphi}_K\tilde{p}_n)$ converges to $\tilde{0}$ for the topology $\sigma(L_R^1,L_R^\infty)$. The first term is dominated by

$$2a\int\varphi_K|\langle g,z'_n\rangle-\langle g,z'\rangle|\,d\mu$$

which tends to 0 by virtue of (2) and Lebesgue's theorem. This contradicts (1) and hence proves the first part of the proposition.

To prove 2) it is enough to note that $S\subset\overline{\bigcup_{n=1}^{\infty}S(n)}$ (= closure in the norm topology) and to use 1). This completes the proof.

Let now ρ be a lifting of M_R^∞. We have:

Theorem 3. – *Let $f\in M_E^\infty[E']$. Then $\rho(f)$ is measurable, that is, $\rho(f)\in\mathscr{L}_E^\infty$.*

Proof: Let $g=\rho(f)$ and let U be the operator corresponding to g as in Corollary 2. We shall show that the function g satisfies the conditions of Theorem 10 in section 2, chapter 6. Condition 10.1) is clearly satisfied. To verify 10.2) we reason as follows:

Fix $K\in\mathscr{B}_0$. Then $\varphi_{\rho(K)}g\in M_E^\infty[E']$, $\rho(\varphi_{\rho(K)}g=\varphi_{\rho(K)}g$ and the operator V corresponding to $\varphi_{\rho(K)}g$ as in Corollary 2 is the operator

$$V:\tilde{f}\to U(\tilde{\varphi}_K\tilde{f}).$$

From jjj) of Corollary 2 (use also the notation of Proposition 1) we deduce that

$$(\varphi_{\rho(K)}g)(X)\subset\overline{V(\{\tilde{f}\in L_R^1|N_1(\tilde{f})\leqslant 1\})}=\overline{\{U(\tilde{\varphi}_K\tilde{f})|\tilde{f}\in L_R^1,N_1(\tilde{f})\leqslant 1\}}=\bar{S}.$$

The proof is now concluded by applying statement 2) of Proposition 1.

Theorem 4. – *The space $L_E^\infty[E']$ can be canonically identified with a subspace of L_E^∞.*

Proof: Let Φ be the mapping of $L_E^\infty[E']$ into L_E^∞ defined by

$$\Phi(\dot{f})=\widetilde{\rho(f)}.$$

It is easy to see that Φ is a well defined linear mapping of $L_E^\infty[E']$ into L_E^∞ (use Theorem 3). Let now $f\in M_E^\infty[E']$. Clearly

$$N_\infty(\widetilde{\rho(f)})=N_\infty(\rho(f)).$$

By Proposition 1, section 4, chapter 6

$$N_\infty(\rho(f))=N_\infty(\dot{f}).$$

Thus

$$N_\infty(\dot{f})=N_\infty(\widetilde{\rho(f)})$$

and Theorem 4 is proved.

We shall conclude this section with the following theorems (the notation used below is that of chapter 6):

Theorem 5. – *Suppose that the Banach space E is reflexive. Let $g: X\to E$ be such that*[1]*: There is $(K_j)_{j\in J}\in\mathscr{Y}$ such that $\varphi_{K_j}\langle g,z'\rangle\in M_R^\infty$ and*

$$\rho(\varphi_{K_j}\langle g,z'\rangle)=\varphi_{\rho(K_j)}\langle g,z'\rangle$$

for all $j\in J$ and $z'\in E'$. Then $g\in\mathscr{L}_E$.

Proof: It is obviously enough to show that $\varphi_{\rho(K_j)}g\in\mathscr{L}_E$ for all $j\in J$. Fix now $j\in J$. From the assumptions of the theorem it follows that $\varphi_{\rho(K_j)}g\in M_E^\infty[E']$ and that $\rho(\varphi_{\rho(K_j)}g)=\varphi_{\rho(K_j)}g$. By Theorem 3, $\varphi_{\rho(K_j)}g$ is measurable. Hence the proof of Theorem 5 is concluded.

From Theorems 3 and 5 we deduce:

Theorem 6. – *Suppose that the Banach space E is reflexive. We have:*

6.1) *If $g\in M_{E'}^\infty[E]$ and $\rho(g)=g$ then $g\in\mathscr{L}_{E'}^\infty$.*

6.2) *If $g\in\mathscr{L}_{E'}^p[E]$ $(1\leq p<+\infty)$ and $\rho[g]=g$ then $g\in\mathscr{L}_{E'}^p$.*

Hence for E reflexive, $L_{E'}^\infty[E]=L_{E'}^\infty$ and $L_{E'}^p[E]=L_{E'}^p$ if $1\leq p<+\infty$.

4. The dual of L_E^1

Let E be a *Banach space*. Let $g\in M_{E'}^\infty[E]$. If $f\in\mathscr{R}_E$, then f is of the form $\sum_{j=1}^n f_j a_j$ with $f_j\in\mathscr{R}$ and $a_j\in E$ $(1\leq j\leq n)$ and hence

$$\langle f,g\rangle=\sum_{j=1}^n f_j\langle a_j,g\rangle$$

[1]) The assumption may be weakened. It is enough to suppose that $(K_j)_{j\in J}$ is a family (not necessarily countable) of elements belonging to $\mathscr{B}$, such that $\sup\{\tilde{K}_j|j\in J\}=\tilde{X}$.

is measurable. Using Definition 2, section 2, chapter 6, we deduce that $\langle f,g\rangle$ is measurable for each $f\in\mathscr{L}_E$.

In the same way we see that if $f\in\mathscr{L}_E$ and if g_1 and g_2 belong to $M_{E'}^{\infty}[E]$ and $g_1\equiv g_2$ (w) then

$$\langle f,g_1\rangle\equiv\langle f,g_2\rangle.$$

If $f\in\mathscr{L}_E^1$ and $g\in M_{E'}^{\infty}[E]$ we have

$$N_1(\langle f,g\rangle)\leqslant N_1(\|f\|\;\|g\|)\leqslant N_1(f)N_\infty(g)<+\infty;$$

in particular we conclude that $\langle f,g\rangle\in\mathscr{L}_R^1$. If we recall that

$$N_\infty(\dot g)=\inf\{N_\infty(h)|h\in M_{E'}^{\infty}[E],\ \dot h=\dot g\}$$

we deduce

(1) $$N_1(\langle f,g\rangle)\leqslant N_1(\tilde f)N_\infty(\dot g).$$

Let now ρ be a lifting of M_R^∞. We have:

Theorem 7. – 1) *Let $u\in(L_E^1)'$. Then there is a unique $g\in M_{E'}^{\infty}[E]$ satisfying:*

1.1) $\rho(g)=g$;

1.2) $u(\tilde f)=\int\langle f(t),g(t)\rangle\,d\mu(t)$ *for all* $\tilde f\in L_E^1$.

Moreover:

1.3) $N_\infty(\dot g)=\|g\|_\infty=\|u\|$.

2) *Conversely if $\dot g\in L_{E'}^{\infty}[E]$ then the mapping $u\colon L_E^1\to R$ defined by*

$$u(\tilde f)=\int\langle f(t),g(t)\rangle\,d\mu(t)\quad \textit{for all}\quad \tilde f\in L_E^1$$

belongs to $(L_E^1)'$.

Proof: 1) Let $z\in E$. The mapping $\tilde f\to u(\tilde f z)$ of L_R^1 into R is linear, continuous and of norm $\leqslant\|u\|\;\|z\|$. There is then $h_z\in M_R^\infty$ such that

$$u(\tilde f z)=\int f(t)\,h_z(t)\,d\mu(t)\quad\text{for all}\quad f\in\mathscr{L}_R^1$$

and

$$N_\infty(h_z)\leqslant\|u\|\;\|z\|.$$

It is clear that for any $x\in E$, $y\in E$ and $\lambda\in R$

$$h_{\lambda x}\equiv\lambda h_x\quad\text{and}\quad h_{x+y}\equiv h_x+h_y.$$

It follows that

$$\rho(h_{\lambda x})=\lambda\rho(h_x)\quad\text{and}\quad\rho(h_{x+y})=\rho(h_x)+\rho(h_y).$$

Moreover

$$\|\rho(h_x)\|_\infty=N_\infty(h_x)\leqslant\|u\|\;\|x\|.$$

This shows that for each $t \in X$, the mapping $x \to \rho(h_x)(t)$ belongs to E'. We shall write

$$\rho(h_x)(t) = \langle x, g(t) \rangle \quad \text{for all} \quad x \in E.$$

Clearly $g \in M_{E'}^{\infty}[E], g$ satisfies 1.1) and

(2) $$\|g\|_{\infty} \leqslant \|u\|.$$

Since

$$u(\tilde{f} z) = \int f(t) \langle z, g(t) \rangle \, d\mu(t)$$

for all $f \in \mathscr{R}$, it follows by linearity that

(3) $$u(\tilde{f}) = \int \langle f(t), g(t) \rangle \, d\mu(t)$$

for all $f \in \mathscr{R}_E$. By approximation we deduce that (3) remains valid for all $\tilde{f} \in L_E^1$ (use (1)).

The uniqueness of g follows from the fact that if g_1 and g_2 belong to $M_{E'}^{\infty}[E]$ and

$$\int \langle f(t), g_1(t) \rangle \, d\mu(t) = \int \langle f(t), g_2(t) \rangle \, d\mu(t)$$

for all $\tilde{f} \in L_E^1$ then $g_1 \equiv g_2$ (w) (use functions of the form $\tilde{f} z$ with $f \in \mathscr{L}_R^1$).

Using (1), (2) and (3) we deduce

$$\|u\| \leqslant N_{\infty}(\dot{g}) \leqslant \|g\|_{\infty} \leqslant \|u\|$$

whence

$$\|u\| = N_{\infty}(\dot{g}) = \|g\|_{\infty}.$$

Since the assertion in 2) follows immediately from (1), the theorem is proved.

By Theorem 7, to each $u \in (L_E^1)'$ corresponds a unique $g \in M_{E'}^{\infty}[E]$ satisfying 1.1), 1.2) and 1.3). Denote by Φ_1 the mapping $u \to \dot{g}$ of $(L_E^1)'$ into $L_{E'}^{\infty}[E]$. Then:

Corollary. – *The mapping Φ_1 is an (isometric) isomorphism of the Banach space $(L_E^1)'$ onto the Banach space $L_{E'}^{\infty}[E]$.*

We showed therefore that $(L_E^1)' = L_{E'}^{\infty}[E]$. If E' is of *countable type*, it follows that $(L_E^1)' = L_{E'}^{\infty}$.

From Theorem 6 we also deduce:

Theorem 8. – *Suppose that the Banach space E is reflexive. Then $(L_E^1)' = L_{E'}^{\infty}$.*

5. The dual of L^p_E $(1<p<+\infty)$

Let E be a *Banach space*. We begin this section with the following proposition:

Proposition 2 (Dinculeanu-Foiaş).– *Let* $1\leqslant p<+\infty$ *and let* $U\in(L^p_E)'$. *There is then* $h\in\mathscr{L}^q_{\mathbf{R}}$ $(1/p+1/q=1)$, $h\geqslant 0$, *such that* $N_q(h)\leqslant\|U\|$ *and*

$$|U(\tilde g)|\leqslant\int\|g(t)\|\,h(t)\,d\mu(t)$$

for all $g\in\mathscr{S}_E$.

Proof: The case $p=1$ is obvious (take $h=\|U\|$); we may therefore assume $1<p<+\infty$.

For every $g\in\mathscr{S}_E$, $\tilde f\to U(\tilde f\tilde g)$ is a continuous linear form on $L^p_{\mathbf{R}}$. Hence there is $u_{\tilde g}\in\mathscr{L}^q_{\mathbf{R}}$ such that

$$U(\tilde f\tilde g)=\int f u_{\tilde g}\,d\mu \quad \text{for all} \quad f\in\mathscr{L}^p_{\mathbf{R}}. \tag{1}$$

Let now $\mathscr{F}$ be the set consisting of all functions of the form

$$\sum_j \varphi_{A_j}|u_{\tilde g_j}| \quad \text{(finite sum)}$$

where $A_j\in\mathscr{B}_0(X,N,\mathscr{R})$, $A_{j'}\cap A_{j''}=\emptyset$ if $j'\neq j''$ and $g_j\in\mathscr{S}_E$, $\|g_j\|_\infty\leqslant 1$. It is easy to see that $\mathscr{F}$ is *directed* for the relation $\leqslant$.

Let now $\beta=\sum_j\varphi_{A_j}|u_{\tilde g_j}|\in\mathscr{F}$. Then, if $f\in\mathscr{L}^p_{\mathbf{R}}$,

$$\int f\Big(\sum_j\varphi_{A_j}|u_{\tilde g_j}|\Big)d\mu=\sum_j\int f\varphi_{A_j}h_j u_{\tilde g_j}\,d\mu$$

where h_j are measurable functions satisfying $|h_j|=1$. By (1)

$$|\textstyle\int f\beta\,d\mu|=\Big|\sum_j U(\tilde f\tilde\varphi_{A_j}\tilde h_j\tilde g_j)\Big|=\Big|U\Big(\sum_j\tilde f\tilde\varphi_{A_j}\tilde h_j\tilde g_j\Big)\Big|$$

$$\leqslant\|U\|\,N_p\Big(\tilde f\Big(\sum_j\varphi_{A_j}h_j g_j\Big)\Big).$$

Since $\Big\|\sum_j\varphi_{A_j}h_j g_j\Big\|_\infty\leqslant 1$ we deduce

$$|\textstyle\int f\beta\,d\mu|\leqslant\|U\|\,N_p(\tilde f)$$

for all $\tilde f\in L^p_{\mathbf{R}}$, whence

$$N_q(\beta)\leqslant\|U\|.$$

By Theorem 4, section 2, chapter 1, the set $\{\tilde v\,|\,v\in\mathscr{F}\}$ has a supremum in L^q, say $\tilde h$ and $N_q(\tilde h)\leqslant\|U\|$.

Let now $g \in \mathscr{S}_E$ and let $g_1 : X \to E$ be defined by

$$g_1(t) = \begin{cases} g(t)/\|g(t)\| & \text{if} \quad g(t) \neq 0, \\ 0 & \text{if} \quad g(t) = 0. \end{cases}$$

Clearly $g_1 \in \mathscr{S}_E$, $\|g_1\|_\infty \leqslant 1$ and $g = \|g\| g_1$. We conclude that

$$|U(\tilde{g})| = |U(\|\tilde{g}\| \tilde{g}_1)| = \int \|g(t)\| u_{\tilde{g}_1}(t) \, d\mu(t) \leqslant \int \|g(t)\| h(t) \, d\mu(t).$$

This proves the proposition.

Before obtaining the dual of L^p_E $(1<p<+\infty)$ we shall make several remarks analogous to those at the beginning of section 4.

Let $1<p<+\infty$ $(1/p+1/q=1)$, let $g \in \mathscr{L}^q_{E'}[E]$ and $f \in \mathscr{R}_E$. Since f is of the form $\sum_{j=1}^{n} f_j a_j$ with $f_j \in \mathscr{R}$ and $a_j \in E$ $(1 \leqslant j \leqslant n)$ we obtain

$$\langle f, g \rangle = \sum_{j=1}^{n} f_j \langle a_j, g \rangle$$

whence $\langle f, g \rangle$ is measurable. Using Definition 2, section 2, chapter 6, we deduce that if $f \in \mathscr{L}_E$ then $\langle f, g \rangle$ is measurable.

In the same way we see that if $f \in \mathscr{L}_E$ and if g_1 and g_2 belong to $\mathscr{L}^q_{E'}[E]$ and $g_1 \equiv g_2$ (w) then

$$\langle f, g_1 \rangle \equiv \langle f, g_2 \rangle.$$

If $f \in \mathscr{L}^p_E$ and $g \in \mathscr{L}^q_{E'}[E]$ we have

$$N_1(\langle f, g \rangle) \leqslant N_1(\|f\| \, \|g\|) \leqslant N_p(f) N_q(g) < +\infty;$$

in particular we conclude that $\langle f, g \rangle \in \mathscr{L}^1_R$. If we recall that

$$N_q(\dot{g}) = \inf\{N_q(h) \mid h \in \mathscr{L}^q_{E'}[E], \dot{h} = \dot{g}\}$$

we deduce

(2) $$N_1(\langle f, g \rangle) \leqslant N_p(\tilde{f}) N_q(\dot{g}).$$

Let now ρ be a lifting of M^∞_R. We have:

Theorem 9. – 1) *Let* $1<p<+\infty$ *and* $U \in (L^p_E)'$. *There is then* $g \in \mathscr{L}^q_{E'}[E]$ $(1/p+1/q=1)$ *uniquely determined almost everywhere satisfying:*

1.1) $\rho[g] = g$;
1.2) $U(\tilde{f}) = \int \langle f(t), g(t) \rangle \, d\mu(t)$ *for all* $\tilde{f} \in L^p_E$.

Moreover:

1.3) $N_q(\dot{g}) = N_q(g) = \|U\|$.

2) *Conversely if* $\dot{g} \in L^q_{E'}[E]$, *then the mapping* $U: L^p_E \to R$ *defined by*

$$U(\tilde{f}) = \int \langle f(t), g(t) \rangle \, d\mu(t) \quad \textit{for all} \quad \tilde{f} \in L^p_E$$

belongs to $(L^p_E)'$.

Proof: 1) Let $U \in (L^p_E)'$. By Proposition 2 there is $h \in \mathscr{L}^q_R$, $h \geqslant 0$ such that $N_q(h) \leqslant \|U\|$ and

$$|U(\tilde{g})| \leqslant \int \|g(t)\| \, h(t) \, d\mu(t)$$

for all $g \in \mathscr{S}_E$. Since $h \in \mathscr{L}^q_R$ there is $(K_j)_{j \in J} \in \mathscr{Y}$ such that $\varphi_{K_j} h \in M^\infty$ for each $j \in J$; we may suppose $\rho(\varphi_{K_j}) = \varphi_{K_j}$ and $\rho(\varphi_{K_j} h) = \varphi_{K_j} h$ for all $j \in J$. For each $j \in J$ let

$$M_j = \|\varphi_{K_j} h\|_\infty .$$

Fix now $j \in J$. Denote by $U^{(j)}$ the mapping

$$\tilde{f} \to U(\tilde{\varphi}_{K_j} \tilde{f})$$

of $\tilde{\mathscr{S}_E}$ into R. Clearly $U^{(j)}$ is linear and

$$|U^{(j)}(\tilde{f})| = |U(\tilde{\varphi}_{K_j} \tilde{f})| \leqslant \int \varphi_{K_j}(t) \|f(t)\| \, h(t) \, d\mu(t)$$
$$\leqslant M_j \int \|f(t)\| \, d\mu(t) = M_j N_1(\tilde{f})$$

for $\tilde{f} \in \tilde{\mathscr{S}_E}$. Hence $U^{(j)}$ *is a linear continuous mapping of* $\tilde{\mathscr{S}_E} \subset L^1_E$ into R. By Theorem 7 there exists $g_j \in M^\infty_{E'}[E]$ such that $\rho(g_j) = g_j$ and

$$U^{(j)}(\tilde{f}) = \int \langle f(t), g_j(t) \rangle \, d\mu(t)$$

for all $f \in \mathscr{S}_E$. Let now $f \in \mathscr{S}_R$ and $x \in E$; we deduce

$$\left| \int f(t) \langle x, g_j(t) \rangle \, d\mu(t) \right| = |U^{(j)}(\tilde{f} x)| \leqslant \|x\| \int \varphi_{K_j}(t) |f(t)| h(t) \, d\mu(t).$$

Since $f \in \mathscr{S}_R$ was arbitrary we conclude

$$(3) \qquad |\langle x, g_j(t) \rangle| \leqslant \|x\| \, \varphi_{K_j}(t) h(t) \quad \text{for all} \quad t \in X.$$

Define now $g = \sum_{j \in J} g_j$. By (3) we have $N_q(g) \leqslant N_q(h)$ and therefore $g \in \mathscr{L}^q_{E'}[E]$. Moreover $\varphi_{K_j} g = g_j$ for all $j \in J$, whence

$$\rho(\varphi_{K_j} \langle z, g \rangle) = \rho(\langle z, g_j \rangle) = \langle z, g_j \rangle = \varphi_{\rho(K_j)} \langle z, g \rangle$$

for all $z \in E$. Hence $\rho[g] = g$ and hence $g \in \mathscr{L}^q_{E'}[E]$ satisfies 1.1).

Let $H = \{j_1, \ldots, j_n\} \subset J$ a finite part and $K_H = K_{j_1} \cup \cdots \cup K_{j_n}$. Then for $f \in \mathscr{S}_E$ we have

$$U(\tilde{\varphi}_{K_H} \tilde{f}) = \sum_{l=1}^{n} U^{(j_l)}(\tilde{f}) = \sum_{l=1}^{n} \int \langle f(t), g_{j_l}(t) \rangle \, d\mu(t)$$
$$= \int \varphi_{K_H}(t) \langle f(t), g(t) \rangle \, d\mu(t) = \int \langle (\varphi_{K_H} f)(t), g(t) \rangle \, d\mu(t).$$

Since $H \subset J$ was arbitrary we deduce

$$U(\tilde{f}) = \int \langle f(t), g(t) \rangle \, d\mu(t)$$

for every $\tilde{f} \in \tilde{\mathscr{S}}_E$ and hence for all $\tilde{f} \in L^p_E$. Thus g satisfies also 1.2).

The assertion that g is uniquely determined almost everywhere is immediate if we use Proposition 3, section 6, chapter 6.

We already know that $N_q(g) \leqslant N_q(h) \leqslant \|U\|$. On the other hand, $\|U\| \leqslant N_q(g)$ (use (2)). Since $\rho[g] = g$ we have $N_q(\dot{g}) = N_q(g)$ (use again Proposition 3, section 6, chapter 6). Thus 1.3) is also proved.

Since the assertion in 2) follows immediately from (2), this completes the proof of the theorem.

By Theorem 9, to each $U \in (L^p_E)'$ corresponds a $g \in \mathscr{L}^q_{E'}[E]$ $(1 < p < +\infty, 1/p + 1/q = 1)$ uniquely determined almost everywhere satisfying 1.1), 1.2) and 1.3). Denote by Φ_p the mapping $U \to \dot{g}$ of $(L^p_E)'$ into $L^q_{E'}[E]$. Then:

Corollary. – *The mapping Φ_p is an (isometric) isomorphism of the Banach space $(L^p_E)'$ $(1 < p < +\infty)$ onto the Banach space $L^q_{E'}[E]$ $(1/p + 1/q = 1)$.*

Let $1 < p < +\infty$ and $1/p + 1/q = 1$. We showed therefore that $(L^p_E)' = L^q_{E'}[E]$. If E' is of *countable type*, it follows that $(L^p_E)' = L^q_{E'}$.

From Theorem 6 we also deduce:

Theorem 10. – *Suppose that the Banach space E is reflexive. Then $(L^p_E)' = L^q_{E'}$.*

6. A theorem of Strassen

Let E be a *Banach space*.

Let $q: E \to R$. We recall that q is *subadditive* if

$$q(x+y) \leqslant q(x) + q(y)$$

for all $x \in E$, $y \in E$. The mapping q is *positively homogeneous* if

$$q(\lambda x) = \lambda q(x)$$

for all $\lambda > 0$ and $x \in E$.

For any subadditive and positively homogeneous mapping $q: E \to R$ we shall write

$$\|q\| = \sup_{\|x\| \leqslant 1} |q(x)|.$$

Then q is continuous if and only if $\|q\| < +\infty$. Moreover, if q is continuous, we have $|q(x)| \leqslant \|q\| \, \|x\|$ for all $x \in E$.

Below we shall call *support function on E* any mapping $q: E \to R$ which is continuous, subadditive and positively homogeneous. We shall denote by $S(E)$ *the set of all support functions on* E; *note that* $E' \subset S(E)$.

If $u \in S(E)$ and $x \in E$, we shall sometimes write $\langle x, u\rangle$ instead of $u(x)$. If $\gamma \in E'$ and $u \in S(E)$ we shall say that γ *is dominated by* u *on* E if

$$\langle x, \gamma\rangle \leqslant \langle x, u\rangle$$

for all $x \in E$.

If $p: t \to p_t$ is a mapping of X into $S(E)$ we denote by $\|p\|$ the mapping $t \to \|p_t\|$, and for each $x \in E$ we denote by $\langle x, p\rangle$ the mapping $t \to \langle x, p_t\rangle$.

Let now ρ *be a lifting of* M_R^∞. We have:

Theorem 11 (Strassen). – *Let* $p: t \to p_t$ *be a mapping of* X *into* $S(E)$. *We suppose that:*

11.1) $N_1(\|p\|) < \infty$ *and* $\|p\|_\infty < \infty$.

11.2) *For each* $x \in E$ *the mapping* $\langle x, p\rangle$ *is measurable.*

11.3) *There is a negligible set* $A \subset X$ *such that*

$$\rho(\langle x, p\rangle)(t) \leqslant \langle x, p_t\rangle$$

for every $x \in E$ *and* $t \in \complement A$.

Denote by q *the element of* $S(E)$ *defined by*

$$\langle x, q\rangle = \int \langle x, p_t\rangle \, d\mu(t) \quad \textit{for} \quad x \in E.$$

Let $x' \in E'$. *Then the following assertions are equivalent:*

a) *The continuous linear form* x' *is dominated by* q.

b) *There exists* $\lambda: t \to \lambda_t$ *in* $M_{E'}^\infty[E]$ *such that* λ_t *is dominated by* p_t *for each* $t \in X$, $\rho(\langle x, \lambda\rangle)(t) = \langle x, \lambda_t\rangle$ *for all* $t \in \complement A$ *and*

$$\langle x, x'\rangle = \int \langle x, \lambda_t\rangle \, d\mu(t) \quad \textit{for all} \quad x \in E.$$

Proof: The implication b)$\Rightarrow$a) is obvious. It remains to prove a)$\Rightarrow$b). By 11.1) there is a set X_0, which is not negligible[1]) and is a countable union of integrable sets, such that $p_t = 0$ for $t \in \complement X_0$. We now note that we may identify E with a subspace of L_E^∞. In fact let θ be the mapping which associates with every $a \in E$ the element $\tilde{\varphi}_{X_0} a$ in L_E^∞. Clearly θ is an isomorphism of the Banach space E onto a subspace of L_E^∞; *in the rest of the proof we identify* E *with this subspace of* L_E^∞.

For every $f \in \mathscr{L}_E^\infty$ the mapping $t \to \langle f(t), p_t\rangle$ of X into R is measurable and integrable (use 11.1) and 11.2). We may then extend q to a mapping $Q: L_E^\infty \to R$ by setting

$$\langle \tilde{f}, Q\rangle = \int \langle f(t), p_t\rangle \, d\mu(t) \quad \text{for} \quad \tilde{f} \in L_E^\infty;$$

---

[1]) Whenever we consider a lifting of M_R^∞ we assume (implicitly or explicitly) that $(X, N, \mathscr{R})$ is strictly localizable and that $N(1) \neq 0$.

clearly $Q \in S(L_E^\infty)$. By the Hahn-Banach theorem, $x' \in E'$ dominated by q on E can be extended to a $\xi' \in (L_E^\infty)'$ dominated by Q on L_E^∞.

For $\tilde{f} \in L_E^1 \cap L_E^\infty$ we have

$$|\langle \tilde{f}, Q\rangle| \leqslant \int |\langle f(t), p_t\rangle| \, d\mu(t) \leqslant N_1(\tilde{f}) N_\infty(\|p\|),$$

whence

$$|\langle \tilde{f}, \xi'\rangle| \leqslant N_1(\tilde{f}) N_\infty(\|p\|).$$

Thus the restriction of ξ' to the linear subspace $L_E^1 \cap L_E^\infty$ of L_E^1 is continuous (for the L_E^1-norm) and of norm $\leqslant N_\infty(\|p\|)$. By Theorem 7 there is $g \in M_{E'}^\infty[E]$ such that $\rho(g) = g$ and

$$\langle \tilde{f}, \xi'\rangle = \int \langle f(t), g(t)\rangle \, d\mu(t)$$

for all $\tilde{f} \in L_E^1 \cap L_E^\infty$. Let now $x \in E$. Then for every $B \in \mathscr{B}_0$, $\tilde{\varphi}_B x \in L_E^1 \cap L_E^\infty$, whence

$$\int \varphi_B(t) \langle x, g(t)\rangle \, d\mu(t) \leqslant \int \varphi_B(t) \langle x, p_t\rangle \, d\mu(t).$$

We deduce

$$\langle x, g(t)\rangle \leqslant \langle x, p_t\rangle$$

almost everywhere on X, whence (use 11.3))

$$\langle x, g(t)\rangle \leqslant \langle x, p_t\rangle$$

for all $t \in \complement A$. Since $x \in E$ was arbitrary it follows that for $t \in \complement A$, $g(t)$ is dominated by p_t. Let now $\lambda \in M_{E'}^\infty[E]$ be defined by

$$\lambda_t = \begin{cases} g(t) & \text{if } t \in \complement A, \\ x_t' & \text{if } t \in A \end{cases}$$

where for $t \in A$, x_t' is any element of E' dominated by p_t. Clearly

$$\langle \tilde{\varphi}_B x, \xi'\rangle = \int \varphi_B(t) \langle x, \lambda_t\rangle \, d\mu(t)$$

for all $B \in \mathscr{B}_0$ and $x \in E$. Since ξ' is dominated by[1]) Q on L_E^∞ and since $N_1(\|\lambda\|) \leqslant N_1(\|p\|) < +\infty$ (note that $\|\lambda_t\| \leqslant \|p_t\|$ for all $t \in X$) we may apply the Lebesgue theorem and we obtain

$$\langle x, x'\rangle = \langle \tilde{\varphi}_{X_0} x, \xi'\rangle = \int \varphi_{X_0}(t) \langle x, \lambda_t\rangle \, d\mu(t);$$

therefore

$$\langle x, x'\rangle = \int \langle x, \lambda_t\rangle \, d\mu(t)$$

for all $x \in E$ and hence the theorem is completely proved.

---

[1]) Note that if $\tilde{f} \in L_R^\infty$ and $x \in E$ then

$$\pm\langle \tilde{f} x, \xi'\rangle \leqslant \langle \pm \tilde{f} x, Q\rangle \leqslant \int |f(t)| \, |\langle x, p_t\rangle| \, d\mu(t);$$

hence

$$|\langle \tilde{f} x, \xi'\rangle| \leqslant \int |f(t)| \, |\langle x, p_t\rangle| \, d\mu(t).$$

Remark. – If E is of *countable type* then it is easy to see that *every* mapping $p: t \to p_t$ of X into $S(E)$ satisfying 11.1) and 11.2) satisfies also 11.3) for a convenient choice of the negligible set A.

7. An application to stochastic processes

We assume in this section that $N(1)=1$, that $T \subset \bar{R}$ *is an interval of* $\bar{R}$ *and that* E *is a completely regular space.*

We recall the following definitions:

(1) A family $(X_t)_{t\in T}$ is a *stochastic process* defined on $(X, N, \mathcal{R})$ with values in E if for each $t \in T$, $X_t \in M_E^\infty$ (see section 5, chapter 4).

(2) Let $(X_t)_{t\in T}$ and $(Y_t)_{t\in T}$ be two stochastic processes defined on $(X, N, \mathcal{R})$ with values in E. The process $(Y_t)_{t\in T}$ is a *modification* of $(X_t)_{t\in T}$ if $Y_t \equiv X_t$ for each $t \in T$.

We now recall the definition of a *separable stochastic process* as given in [96].

Let $(Y_t)_{t\in T}$ be a stochastic process defined on $(X, N, \mathcal{R})$ with values in E. Let I be a subset of T, K a compact Baire set in E. We define

$$V(I,K) = \{x \mid Y_t(x) \in K \quad \text{for all} \quad t \in I\}.$$

If I is countable, then clearly $V(I,K) \in \mathcal{B}$ and

$$\mu(V(I,K)) = \inf_{A \subset I,\, A \text{ finite}} \mu(V(A,K)).$$

Definition 1. – *Let* $(Y_t)_{t\in T}$ *be a stochastic process defined on* $(X, N, \mathcal{R})$ *with values in* E. *We say that the process is separable if for every open interval* $I \subset T$ *and every compact Baire set* $K \subset E$:

i) $V(I,K) \in \mathcal{B}$;
ii) $\mu(V(I,K)) = \inf_{A \subset I,\, A \text{ finite}} \mu(V(A,K))$.

Let now ρ *be a lifting of* M_R^∞. We may then state the following:

Theorem 12. – *Let* $(X_t)_{t\in T}$ *be a stochastic process defined on* $(X, N, \mathcal{R})$ *with values in* E. *The process* $(Y_t)_{t\in T}$ *defined by*

$$Y_t = \rho_E(X_t) \quad \textit{for each} \quad t \in T,$$

is a separable modification of $(X_t)_{t\in T}$.

Proof: Let $K \subset E$ be a compact set. Then (use Remark 3) following Definition 3 in section 5, chapter 4)[1]) the set $Y_t^{-1}(K)$ is measurable and

$$\rho(Y_t^{-1}(K)) \subset Y_t^{-1}(K)$$

---

[1]) Note that we do not have to suppose that the compact K is a Baire set.

for every $t \in T$. Hence if $A \subset T$ is finite we have

$$\rho(V(A,K)) = \rho\Big(\bigcap_{t\in A} Y_t^{-1}(K)\Big) = \bigcap_{t\in A} \rho(Y_t^{-1}(K)) \subset \bigcap_{t\in A} Y_t^{-1}(K) = V(A,K).$$

Let now $I \subset T$. Since the set $(V(A,K))_{A\subset I,\, A \text{ finite}}$ is directed (for $\supset$) and has for intersection $V(I,K)$ we deduce (use Theorem 3, section 3, chapter 3) $V(I,K) \in \mathscr{B}$ and

$$\mu(V(I,K)) = \inf_{A\subset I,\, A \text{ finite}} \mu(V(A,K)).$$

This proves Theorem 12.

CHAPTER VIII

Strong liftings

Throughout this chapter, X will be either a compact or a locally compact space and $\mu \neq 0$ a positive Radon measure on X. We shall write $M_R^\infty(X,\mu)$, or simply M_R^∞, instead of $M_R^\infty(X,\bar{N},\mathscr{K}_R(X))$ (where $\bar{N}=\mu^\bullet$ is of course the essential upper integral associated with μ). Similarly we shall write $L_R^\infty(X,\mu)$, or simply L_R^∞, instead of $L_R^\infty(X,\bar{N},\mathscr{K}_R(X))$ (see sections 7 and 9, chapter 1).

In this and in the next chapters we shall often write $\mu^\bullet$ instead of $\bar{N}$. We shall also write $\mu^\bullet(A)$ instead of $\bar{N}(\varphi_A)$ if $A \subset X$.

We recall that if X is a locally compact space and μ a positive Radon measure on X, then $(X,\mu^\bullet,\mathscr{K}_R(X))$ is strictly localizable. The equivalence relations "$\equiv$" considered here are of course taken "with respect to $\mu^\bullet$".

We recall that for any locally compact space X, $C_R(X)$ is the algebra of all real-valued continuous functions on X and that $C_R^b(X)=\{f\in C_R(X) \mid f \text{ bounded}\}$.

In this chapter we shall use, without further explanation, various definitions and results from Bourbaki's *Intégration*.

1. The notion of strong lifting

Let X be a *locally compact space* and $\mu \neq 0$ a *positive Radon measure on* X.

Definition 1. – *Assume that* $\operatorname{Supp}\mu = X$ *and let* ρ *be a linear lifting (respectively, a lifting) of* $M_R^\infty(X,\mu)$. *We say that* ρ *is a strong linear lifting (respectively, a strong lifting) of* $M_R^\infty(X,\mu)$ *if*[1])

$$\rho(f)=f \qquad \textit{for all} \qquad f\in C_R^b(X).$$

---

[1]) Note that whenever we say that ρ is a strong linear lifting or a strong lifting of $M_R^\infty(X,\mu)$ we assume (implicitly or explicitly) that $\operatorname{Supp}\mu = X$.

We say that *the couple* (X,μ) *has the strong lifting property* if there is a strong lifting of $M_R^\infty(X,\mu)$.

In the theorem below, the notation $\mathscr{V}(x)$ (= the set of all neighborhoods of $x \in X$) and the term fundamental system of neighborhoods of x *refer to the* (locally compact) *topology* T *of* X.

From Proposition 3 and Theorem 3 in section 5, chapter 5, we immediately obtain:

Theorem 1. – *Assume that* $\operatorname{Supp}\mu = X$ *and let* ρ *be a lifting of* $M_R^\infty(X,\mu)$. *Then the following assertions are equivalent:*

i) ρ *is a strong lifting.*

ii) $\rho(A) \subset D(A,T)$ *for every* $A \subset X$ *which is* $\mu^\bullet$*-measurable.*

iii) $U \subset \rho(U)$ *for every* $U \subset X$ *open* (i.e., ρ *is compatible with* T).

iv) $\rho(F) \subset F$ *for every* $F \subset X$ *closed.*

v) $\rho(f) = f$ *for every* $f \in \mathscr{K}_R(X)$.

vi) *For each* $x \in X$,

$$\mathscr{V}_\rho(x) = \{\rho(V) \mid V \in \mathscr{V}(x) \text{ and } V \ \mu^\bullet\text{-measurable}\}$$

is a fundamental system of neighborhoods of X.

Let us note that if $\gamma: M_R^\infty \to M_R^\infty$ is a *strong linear lifting* of M_R^∞ then for every open set $U \subset X$

$$\varphi_U \leqslant \gamma(\varphi_U).$$

In fact, observe that $\varphi_U = \sup\{h \in C_R^b(X) | 0 \leqslant h \leqslant \varphi_U\}$ and that $\gamma(\varphi_U) \geqslant \gamma(h) = h$ for each $h \in C_R^b(X)$ with $0 \leqslant h \leqslant \varphi_U$.

We denote by $\mathscr{M}_R(X,\mu)$ or simply $\mathscr{M}_R$ since there can be no ambiguity, the algebra of all *locally bounded* (i.e. bounded on every compact) $\mu^\bullet$*-measurable* functions on X to R. Note that $C_R(X)$ is a subalgebra of $\mathscr{M}_R$. A mapping $\rho: \mathscr{M}_R \to \mathscr{M}_R$ verifying properties (I)–(VI) in section 1, chapter 3, will be called a *lifting* of $\mathscr{M}_R$. A *strong lifting* of $\mathscr{M}_R$ is a lifting ρ of $\mathscr{M}_R$ such that $\rho(f) = f$ for all $f \in C_R(X)$.

Theorem 2. – *The following assertions are equivalent:*

j) *There is a strong linear lifting of* M_R^∞.

jj) *There is a strong lifting of* M_R^∞.

jjj) *There is a strong lifting of* $\mathscr{M}_R$.

Proof: j)$\Rightarrow$jj). Let $\gamma: M_R^\infty \to M_R^\infty$ be a strong linear lifting of M_R^∞. Let θ and θ' be the lower and upper densities associated with γ (see the end of section 1, chapter 3). Let now $\mathscr{G}$ $(= \mathscr{G}_{M_R^\infty})$ be the set of all linear liftings of M_R^∞ and $\mathscr{G}(\gamma)$ the set of all $\sigma \in \mathscr{G}$ satisfying the inequalities

$$\varphi_{\theta(A)} \leqslant \sigma(\varphi_A) \leqslant \varphi_{\theta'(A)}$$

for every $A \in \mathscr{B}$ (see section 2, chapter 3). By Theorem 2, section 2, chapter 3, there is a lifting ρ of M_R^∞ belonging to the set $\mathscr{G}(\gamma)$. Let now $U \subset X$ be an open set. By the remark made above, $\varphi_U \leqslant \gamma(\varphi_U)$ and thus $\varphi_U \leqslant \varphi_{\theta(U)}$. We deduce $\varphi_U \leqslant \rho(\varphi_U)$, whence $U \subset \rho(U)$. By Theorem 1, ρ is a strong lifting of M_R^∞.

jj)$\Rightarrow$jjj). Let $\rho: M_R^\infty \to M_R^\infty$ be a strong lifting. Let $f \in \mathscr{M}_R$ and let $t \in X$. Let $V \in \mathscr{V}(t)$ be open and relatively compact. We define

$$\rho'(f)(t) = \rho(\varphi_V f)(t)$$

(note that $\varphi_V f \in M_R^\infty$). If $U \in \mathscr{V}(t)$ is another open relatively compact set, then $\rho(\varphi_U f)(t) = \rho(\varphi_V f)(t)$. In fact, let $W = U \cap V$; since

$$\begin{aligned}\rho(\varphi_U f)(t) &= \rho(\varphi_W)(t)\rho(\varphi_U f)(t) = \rho(\varphi_W \varphi_U f)(t) \\ &= \rho(\varphi_W \varphi_V f)(t) = \rho(\varphi_W)(t)\rho(\varphi_V f)(t) = \rho(\varphi_V f)(t).\end{aligned}$$

We deduce that $\rho'(f)(t)$ is well defined, for $f \in \mathscr{M}_R$ and $t \in X$. To see that $\rho'(f)$ belongs to $\mathscr{M}_R$ for each $f \in \mathscr{M}_R$, let $K \subset X$ be a compact set and let $U \supset K$ be open and relatively compact. For $t \in K$ we have $\rho'(f)(t) = \rho(\varphi_U f)(t)$; this shows that $\varphi_K \rho'(f)$ belongs to M_R^∞. Since K is arbitrary we deduce that $\rho'(f) \in \mathscr{M}_R$. Thus ρ' is a mapping of $\mathscr{M}_R$ into $\mathscr{M}_R$. It is easy to see that this mapping is a strong lifting of $\mathscr{M}_R$. To verify (I), for instance, we may reason as follows:

Let $f \in \mathscr{M}_R$ and let U be open and relatively compact. Then for $t \in U$, $\rho'(f)(t) = \rho(\varphi_U f)(t)$, whence

$$\varphi_U \rho'(f) = \rho(\varphi_U f) \equiv \varphi_U f.$$

Since U was arbitrary we deduce $\rho'(f) \equiv f$. The proof of the properties (II)–(VI) is immediate and is left to the reader. Let now $f \in C_R(X)$ and $t \in X$. Let $V \in \mathscr{V}(t)$ be open, relatively compact, and let $g \in \mathscr{K}_R(X)$ be such that $g(u) = 1$ for $u \in V$. Then

$$\begin{aligned}\rho'(f)(t) &= \rho(\varphi_V f)(t) = \rho(\varphi_V g f)(t) = \rho(\varphi_V)(t)\rho(g f)(t) \\ &= \rho(g f)(t) = (g f)(t) = f(t)\end{aligned}$$

and the assertion is proved. Thus $\rho': \mathscr{M}_R \to \mathscr{M}_R$ is a strong lifting of $\mathscr{M}_R$.

jjj)$\Rightarrow$j). Let $\rho: \mathscr{M}_R \to \mathscr{M}_R$ be a strong lifting of $\mathscr{M}_R$. Then ρ maps M_R^∞ into M_R^∞ (since ρ is positive and $\rho(1) = 1$) and hence the restriction of ρ to M_R^∞ is a strong lifting of M_R^∞. This completes the proof of Theorem 2.

The definition of ρ' shows that $\rho' | M_R^\infty = \rho$. In fact, let $f \in M_R^\infty$ and $t \in X$. Let $V \in \mathscr{V}(t)$ be open and relatively compact. Then (use iii) of Theorem 1)

$$\rho'(f)(t) = \rho(\varphi_V f)(t) = \rho(\varphi_V)(t)\rho(f)(t) = \varphi_{\rho(V)}(t)\rho(f)(t) = \rho(f)(t).$$

For further reference we state here

Proposition 1. – *Let $\mathscr{A}$ be a closed subalgebra of $L^\infty_R(X,\mu)$, containing $\tilde{1}$, and let χ be a character of $\mathscr{A}$. There is then a character θ of $L^\infty_R(X,\mu)$ satisfying $\theta|\mathscr{A}=\chi$.*

One way of proving the above assertion is as follows: Let $\mathscr{A}_C$ be the (closed) algebra of all objects[1]) $\tilde{f}+i\tilde{g}$, where $\tilde{f}$ and $\tilde{g}$ belong to $L^\infty_R(X,\mu)$. The character χ can clearly be extended to $\mathscr{A}_C$. The Šilov boundary of $\mathscr{A}_C$ coincides with the space of characters of $\mathscr{A}_C$. Whence, every character of $\mathscr{A}_C$ can be extended (see [101], p. 195) to a character of $L^\infty_C(X,\mu)$. We conclude that every character of $\mathscr{A}$ can be extended to a character of $L^\infty_R(X,\mu)$.

Note that an element of $L^\infty_R(X,\mu)$, considered as an equivalence class, contains at most one function belonging to $C^b_R(X)$ (recall that we suppose $\operatorname{Supp}\mu=X$). Hence for every character χ of $C^b_R(X)$ there exists a character θ of $L^\infty_R(X,\mu)$ such that $\theta(\tilde{f})=\chi(f)$ for all $f\in C^b_R(X)$. In particular, for every $t\in X$, there exists a character χ_t of $L^\infty_R(X,\mu)$ satisfying $\chi_t(\tilde{f})=f(t)$ for all $f\in C^b_R(X)$.

Let now Y be a *locally compact subspace* of X. We recall that μ_Y (= the measure induced on Y by μ) is defined by the following formulas:

$$\int g\,d\mu_Y=\int g'\,d\mu \quad \text{for each} \quad g\in\mathscr{K}_R(Y),$$

where g' is the function on X to R equal to g on Y and to 0 on $\complement Y$.

We note that if K_1, K_2 are locally compact subspaces of X and $H\subset K_1\subset K_2$, then $\mu^\bullet_{K_2}(H)=\mu^\bullet_{K_1}(H)=\mu^\bullet(H)$ (see [15], chapter 5, p. 82).

We recall that a *family* $(A_j)_{j\in J}$ of subsets of X is said to be *locally countable* if, for each $t\in X$, there is a neighborhood V of t such that the set $\{j\in J|A_j\cap V\neq\varnothing\}$ is countable.

Definition 2. – *We denote by $\mathscr{C}(X,\mu)$ the set of all locally countable families $(K_j)_{j\in J}$ having the following properties:*

(a) *K_j is compact and $\mu(K_j)>0$ for each $j\in J$.*

(b) *$K_{j'}\cap K_{j''}=\varnothing$ if $j'\neq j''$.*

(c) *The set $X-\bigcup_{j\in J}K_j$ is $\mu^\bullet$-negligible (that is, locally μ-negligible).*

It is well known that there are locally countable families $(K_j)_{j\in J}$ satisfying (a), (b), (c) (see [15], chapters 1–4, p. 190–191). Thus the set $\mathscr{C}(X,\mu)$ is *non-void.*

We shall need later the following:

Remarks. – 1) Let $K\subset X$ be a closed set. There is then a closed $K'\subset K$ such that:

α) $\mu^\bullet(K-K')=0$;

β) $\operatorname{Supp}\mu_{K'}=K'$.

---

[1]) The Banach algebra $L^\infty_C(X,\mu)$ (where C is the field of complex numbers) can be defined in the same way as $L^\infty_R(X,\mu)$.

In fact, let $K'=\operatorname{Supp}\mu_K$. Then $K'\subset K$, $\operatorname{Supp}\mu_{K'}\subset K'$ and

$$K-\operatorname{Supp}\mu_{K'}=(K-K')\cup(K'-\operatorname{Supp}\mu_{K'}).$$

But

$$\mu_K^\bullet(K-\operatorname{Supp}\mu_{K'})\leqslant\mu_K^\bullet(K-K')+\mu_K^\bullet(K'-\operatorname{Supp}\mu_{K'})=0+\mu_{K'}^\bullet(K'-\operatorname{Supp}\mu_{K'})=0.$$

Since $K-\operatorname{Supp}\mu_{K'}$ is an open set in K we deduce

$$K-\operatorname{Supp}\mu_{K'}\subset K-\operatorname{Supp}\mu_K,$$

whence

$$\operatorname{Supp}\mu_K(=K')\subset\operatorname{Supp}\mu_{K'}.$$

It follows that

$$\operatorname{Supp}\mu_{K'}=\operatorname{Supp}\mu_K=K'.$$

2) Let $(K_j)_{j\in J}\in\mathscr{C}(X,\mu)$ and for each $j\in J$ let $K'_j\subset K_j$ be a compact set such that:

α') $\mu(K_j-K'_j)=0$;

β') $\operatorname{Supp}\mu_{K'_j}=K'_j$.

Then $(K'_j)_{j\in J}\in\mathscr{C}(X,\mu)$.

In fact let $t\in X$ and let V be a neighborhood of t such that $\{j\in J|K_j\cap V\neq\varnothing\}$ is countable. As $K'_j\cap V\neq\varnothing$ implies $K_j\cap V\neq\varnothing$, the set $\{j\in J|K'_j\cap V\neq\varnothing\}$ is countable and hence the family $(K'_j)_{j\in J}$ is locally countable. Since $(K'_j)_{j\in J}$ is a family of disjoint compact parts of X and since $\mu(K'_j)=\mu(K_j)>0$ for each $j\in J$, $(K'_j)_{j\in J}$ satisfies (a) and (b). To verify (c), let $K\subset X$ be compact. Note that $J_K=\{j\in J|K_j\cap K\neq\varnothing\}$ is countable (since the family $(K_j)_{j\in J}$ is locally countable). But

$$K\cap\Big(X-\bigcup_{j\in J}K'_j\Big)\subset\Big(K\cap\Big(X-\bigcup_{j\in J}K_j\Big)\Big)\cup\Big(\bigcup_{j\in J_K}K\cap(K_j-K'_j)\Big).$$

We deduce that $\mu\Big(K\cap\Big(X-\bigcup_{j\in J}K'_j\Big)\Big)=0$ and this completes the proof.

Proposition 2. – *Let X be a locally compact space and μ a positive Radon measure on X with* $\operatorname{Supp}\mu=X$.

2.1) *Suppose that (X,μ) has the strong lifting property and let $K\subset X$ be a nonvoid compact such that* $\operatorname{Supp}\mu_K=K$. *Then (K,μ_K) has the strong lifting property.*

2.2) *Conversely, let $(K_j)_{j\in J}\in\mathscr{C}(X,\mu)$ be such that* $\operatorname{Supp}\mu_{K_j}=K_j$ *for each $j\in J$ and suppose that (K_j,μ_{K_j}) has the strong lifting property for every $j\in J$. Then (X,μ) has the strong lifting property.*

Proof: 2.1) Let $\rho: M_R^\infty(X,\mu)\to M_R^\infty(X,\mu)$ be a strong lifting of $M_R^\infty(X,\mu)$. For each $t\in K-\rho(K)$ (note that $\rho(K)\subset K$ by Theorem 1) let χ_t be a character of $L_R^\infty(K,\mu_K)$ such that $\chi_t(\tilde{f})=f(t)$ for $f\in C_R(K)$ (use Proposition 1). For $f\in M_R^\infty(K,\mu_K)$ define $f': X\to R$ by

$$f'(t)=\begin{cases} f(t) & \text{if } t\in K,\\ 0 & \text{if } t\notin K.\end{cases}$$

The mapping $f\to f'$ is a representation of $M_R^\infty(K,\mu_K)$ into $M_R^\infty(X,\mu)$.

Define now

$$\rho'(f)(t)=\begin{cases}\rho(f')(t) & \text{if} \quad t\in\rho(K),\\ \chi_t(\tilde{f}) & \text{if} \quad t\in K-\rho(K).\end{cases}$$

Then $\rho': f\to\rho'(f)$ is a representation of $M_R^\infty(K,\mu_K)$ into $M_R^\infty(K,\mu_K)$. It is easily seen the ρ' is a lifting of $M_R^\infty(K,\mu_K)$. To verify that ρ' is in fact a strong lifting of $M_R^\infty(K,\mu_K)$ let $f\in C_R(K)$ and let $h\in C_R^b(X)$ be such that $h|K=f$. Then for $t\in\rho(K)$,

$$\begin{aligned}\rho'(f)(t)&=\rho(\varphi_{\rho(K)})(t)\rho(f')(t)=\rho(\varphi_{\rho(K)}f')(t)=\rho(\varphi_{\rho(K)}h)(t)\\ &=\varphi_{\rho(K)}(t)\rho(h)(t)=h(t)=f(t);\end{aligned}$$

for $t\in K-\rho(K)$,

$$\rho'(f)(t)=\chi_t(\tilde{f})=f(t).$$

Thus $\rho'(f)=f$ and hence ρ' is a strong lifting of $M_R^\infty(K,\mu_K)$.

2.2) For each $j\in J$ let $\rho_j: M_R^\infty(K_j,\mu_{K_j})\to M_R^\infty(K_j,\mu_{K_j})$ be a strong lifting of $M_R^\infty(K_j,\mu_{K_j})$. For every $t\in X-\bigcup_{j\in J}K_j$ let χ_t be a character of $L_R^\infty(X,\mu)$ such that $\chi_t(\tilde{f})=f(t)$ for $f\in C_R^b(X)$ (use Proposition 1). Let now $f\in M_R^\infty(X,\mu)$. Then $f|K_j\in M_R^\infty(K_j,\mu_{K_j})$ for each $j\in J$ and hence we may define

$$\rho(f)(t)=\begin{cases}\rho_j(f|K_j)(t) & \text{if} \quad t\in K_j,\\ \chi_t(\tilde{f}) & \text{if} \quad t\in X-\bigcup_{j\in J}K_j.\end{cases}$$

It is easy to see that $\rho: f\to\rho(f)$ is a lifting of $M_R^\infty(X,\mu)$. Moreover if $f\in C_R^b(X)$, then $f|K_j\in C_R^b(K_j)$ and hence $\rho_j(f|K_j)=f|K_j$ for each $j\in J$. We deduce that $\rho(f)=f$, that is, ρ is a strong lifting of $M_R^\infty(X,\mu)$. This completes the proof of Proposition 2.

2. Further results concerning strong liftings. Examples

In what follows we consider $M_R^\infty(X,\mu)$ as a Banach algebra endowed with the supremum norm $f\to\|f\|_\infty$.

As we shall see in section 4 we have the following:

Theorem 3. – *Let X be a locally compact metrizable space and μ a positive Radon measure on X with* $\operatorname{Supp}\mu=X$. *Then (X,μ) has the strong lifting property.*

Let again X be a *locally compact space* and $\mu\neq0$ *a positive Radon measure on X.*

Let $\mathscr{A}\subset M_R^\infty(X,\mu)$ be a subalgebra and suppose that $1\in\mathscr{A}$ and $\mathscr{N}^\infty\subset\mathscr{A}$. A mapping $\rho:\mathscr{A}\to\mathscr{A}$ satisfying (I)–(VI) of section 1, chapter 3 will be called a *lifting of $\mathscr{A}$*.

Definition 3. – *Let $\mathscr{H} \subset \mathscr{A}$. A lifting ρ of $\mathscr{A}$ is called strong with respect to $\mathscr{H}$ if*

$$\rho(f) = f \quad \text{for all} \quad f \in \mathscr{H}.$$

Remark. – Suppose that $\mathscr{A}$ is closed and that ρ is a lifting of $\mathscr{A}$. Note that if $\mathscr{H}_1$ is the closed algebra spanned by 1 and $\mathscr{H}$, then ρ is strong with respect to $\mathscr{H}_1$ if and only if ρ is strong with respect to $\mathscr{H}$.

It is enough to note that ρ has the "continuity property"

$$\|\rho(f)\|_\infty \leqslant \|f\|_\infty \quad \text{for all} \quad f \in \mathscr{A}.$$

We shall now introduce the[1])

Definition 4. – *A subalgebra $\mathscr{A}$ of $M_R^\infty(X,\mu)$ has the property $(\mathscr{E})$ if:*

i) $1 \in \mathscr{A}$, $\mathscr{N}^\infty \subset \mathscr{A}$.

ii) *For every lifting ρ of $\mathscr{A}$ there is a lifting ρ' of $M_R^\infty(X,\mu)$ satisfying $\rho'|\mathscr{A} = \rho$.*

We may now state and prove the following:

Proposition 3. – *Let X be a locally compact space and μ a positive Radon measure on X with* $\operatorname{Supp}\mu = X$. *Let $\mathscr{A} \subset M_R^\infty(X,\mu)$ be an algebra having the property $(\mathscr{E})$ and let $\mathscr{H} \subset \mathscr{A} \cap C_R^b(X)$. Suppose that:*

a) *There is a lifting ρ of $\mathscr{A}$ which is strong with respect to $\mathscr{H}$.*

Let $\mathscr{D} \subset C_R^b(X)$, $\mathscr{D} \ni 1$ be a countable set. There is then a lifting of $M_R^\infty(X,\mu)$ which is strong with respect to the closed algebra spanned by $\mathscr{H} \cup \mathscr{D}$.

Proof: Since $\mathscr{A}$ has the property $(\mathscr{E})$, there is a lifting ρ_1 of $M_R^\infty(X,\mu)$ satisfying $\rho_1|\mathscr{A} = \rho$. Let $A \subset X$ be a $\mu^\bullet$-negligible (that is, locally μ-negligible) set such that

$$\rho_1(f)(t) = f(t) \quad \text{for all} \quad f \in \mathscr{D} \quad \text{and} \quad t \in \complement A.$$

For each $t \in \complement A$ let χ_t be a character of $L_R^\infty(X,\mu)$ such that $\chi_t(\tilde{f}) = f(t)$ for all $f \in C_R^b(X)$. Define now ρ_2 on $M_R^\infty(X,\mu)$ by the equations:

$$\rho_2(f)(t) = \begin{cases} \rho_1(f)(t) & \text{for} \quad t \in \complement A, \\ \chi_t(\tilde{f}) & \text{for} \quad t \in A. \end{cases}$$

It is then clear that ρ_2 is a lifting of $M_R^\infty(X,\mu)$ such that $\rho_2(f) = f$ for all $f \in \mathscr{H} \cup \mathscr{D}$. By the Remark following Definition 3, ρ_2 is then strong with respect to the closed algebra spanned by $\mathscr{H} \cup \mathscr{D}$ and hence the proposition is proved.

We shall also need the following:

---

[1]) In connection with this definition see also Theorem 4, section 2, chapter 4.

Proposition 4. – *Let* $\mathscr{A} \subset M_R^\infty(X,\mu)$ *be a subalgebra containing* 1 *and closed under pointwise convergence of bounded sequences. Then*

$$\mathscr{A}_1 = \{f+g \mid f \in \mathscr{A},\ g \in \mathscr{N}^\infty\} \quad (=\mathscr{A}+\mathscr{N}^\infty)$$

is an algebra closed under pointwise convergence of bounded sequences[1]).

Proof: It is clear that $\mathscr{A}_1$ is an algebra (since $\mathscr{N}^\infty$ is an ideal in $M_R^\infty(X,\mu)$). Let now (u_n) be a bounded pointwise convergent sequence of functions in $\mathscr{A}_1$. Then $u_n = f_n + g_n$ with $f_n \in \mathscr{A}$ and $g_n \in \mathscr{N}^\infty$ for each n. Let

$$L = \sup_n \|f_n + g_n\|,$$

let C be the set of all $t \in X$ such that

$$\lim_n f_n(t) \quad \text{exists and belongs to} \quad [-L, L],$$

and let h be the pointwise limit of the sequence $(\varphi_C f_n)$. Then $\varphi_C \in \mathscr{A}$ (see Proposition 2, section 1, chapter 2) and hence $h \in \mathscr{A}$. The set $X-C$ is clearly $\mu^\bullet$-negligible. For each n we have:

$$u_n = f_n + g_n = \varphi_C f_n + (\varphi_{X-C} f_n + g_n) = \varphi_C f_n + g'_n$$

(where of course $g'_n = \varphi_{X-C} f_n + g_n \in \mathscr{N}^\infty$). Since $\lim_n \varphi_C(t) f_n(t) = h(t)$ for all $t \in X$, it follows that $\lim_n g'_n(t) = g'(t)$ exists for all $t \in X$ and that $g' \in \mathscr{N}^\infty$. Thus (u_n) converges pointwise to $h + g' \in \mathscr{A}_1$ and the proposition is proved.

Let now X be a compact space, μ a positive Radon measure on X. Let Y be a compact space and $\pi: X \to Y$ a continuous mapping. Let $\nu = \pi(\mu)$. We recall the following well-known facts (see [15], chapter 5, pp. 70–71):

a) A set $A \subset Y$ is $\nu^\bullet$-negligible if and only if $\pi^{-1}(A)$ is $\mu^\bullet$-negligible;

b) $\operatorname{Supp}(\pi(\mu)) = \pi(\operatorname{Supp}\mu)$;

c) A function $f: Y \to R$ is $\nu^\bullet$-measurable if and only if $f \circ \pi$ is $\mu^\bullet$-measurable.

Using Proposition 4 we easily obtain:

Proposition 5.– *Let* X *be a compact space,* μ *a positive Radon measure on* X. *Let* Y *be another compact space,* $\pi: X \to Y$ *a continuous mapping and* $\nu = \pi(\mu)$. *Then*

$$\mathscr{A} = \{f \circ \pi + g \mid f \in M_R^\infty(Y,\nu),\ g \in \mathscr{N}^\infty\}$$

is an admissible subalgebra of $M_R^\infty(X,\mu)$ *and therefore has the property* $(\mathscr{E})$.

---

[1]) Note that if $\mu^\bullet(1) < +\infty$ then $\mathscr{A}_1$ is an admissible subalgebra. In particular $\mathscr{A}_1$ is admissible if X is compact.

Proof: It is enough to note that the set

$$M_R^\infty(Y,\nu)\circ\pi = \{f\circ\pi \mid f\in M_R^\infty(Y,\nu)\}$$

is a subalgebra of $M_R^\infty(X,\mu)$ containing 1 and closed under pointwise convergence of bounded sequences.

For further reference we shall also give (we use here the notations of Proposition 5):

Proposition 6. – *Let ρ be a lifting of $M_R^\infty(Y,\nu)$. If for each $h=f\circ\pi+g\in\mathscr{A}$ (with $f\in M_R^\infty(Y,\nu)$ and $g\in\mathscr{N}^\infty$) we write*

$$\omega(h) = \rho(f)\circ\pi,$$

then ω is a (well defined) lifting of $\mathscr{A}$. Moreover, if ρ is strong with respect to $\mathscr{U}\subset M_R^\infty(Y,\nu)$, then ω is strong with respect to $\mathscr{U}\circ\pi=\{f\circ\pi \mid f\in\mathscr{U}\}$.

Proof: Let $h\in\mathscr{A}$; there is then $f_1\in M_R^\infty(Y,\nu)$ such that $h\equiv f_1\circ\pi$. If $h\equiv f_2\circ\pi$ for some other function $f_2\in M_R^\infty(Y,\nu)$ then $f_1\circ\pi\equiv f_2\circ\pi \Rightarrow f_1\equiv f_2\Rightarrow\rho(f_1)=\rho(f_2)$. Hence ω is defined unambiguously on $\mathscr{A}$ by $\omega(h)=\rho(f_1)\circ\pi$. It is then easy to see that ω is a *lifting* of $\mathscr{A}$.

Moreover if $\rho(f)=f$ for $f\in\mathscr{U}$ then

$$\omega(f\circ\pi)=\rho(f)\circ\pi=f\circ\pi.$$

Hence the proposition is proved.

Note that

$$\omega(\mathscr{A})\subset\{f\circ\pi \mid f\in M_R^\infty(Y,\nu)\}.$$

Corollary. – *Let $X=S\times T$ where S and T are compact spaces and let μ be a positive Radon measure on X with* $\operatorname{Supp}\mu=X$. *Suppose that:*

1) *T is metrizable;*
2) *$(S, pr_S(\mu))$ has the strong lifting property.*

Then (X,μ) has the strong lifting property.

Proof: Let $\mathscr{D}_T\subset C_R(T)$ be a countable set containing 1 and dense in $C_R(T)$ and let

$$\mathscr{D}=\{h\circ pr_T \mid h\in\mathscr{D}_T\}.$$

Let $\mathscr{A}$ be the algebra

$$\{f\circ pr_S+g \mid f\in M_R^\infty(S,pr_S(\mu)),\ g\in\mathscr{N}^\infty\}.$$

By Proposition 5, $\mathscr{A}$ is admissible, whence it has the property $(\mathscr{E})$. Let

$$\mathscr{H}=\{h\circ pr_S \mid h\in C_R(S)\}.$$

Then the closed algebra spanned by $\mathscr{H}\cup\mathscr{D}$ coincides with $C_R(X)$. By Proposition 6, there exists a lifting ω of $\mathscr{A}$, which is strong with respect to $\mathscr{H}$. By Proposition 3, there is then a lifting of $M_R^\infty(X,\mu)$

which is strong with respect to $C_R(X)$. Whence (X,μ) has the strong lifting property.

The following technical result will be used later:

Proposition 7. – *Let $\mathscr{A} \subset M_R^\infty(X,\mu)$ be a subalgebra containing 1 and $\mathscr{N}^\infty$ and closed under pointwise convergence of bounded sequences. Let $\mathscr{D} \subset M_R^\infty(X,\mu)$ be a vector space. Let ω be a lifting*[1]) *of $\mathscr{A}$. Suppose that:*

$$(*) \qquad \varphi_A \in \mathscr{A},\ \omega(A)=A,\ g\in\mathscr{D} \ \text{ and } \ \varphi_A g \equiv 0 \Rightarrow \varphi_A g = 0.$$

Let $\mathscr{S}_\omega$ be the algebra (over R) consisting of all the functions f of the form $f=\sum_{i=1}^{n}\lambda_i\varphi_{Ai}$, where $\lambda_i \in R$, $\varphi_{A_i}\in\mathscr{A}$ and $\omega(A_i)=A_i$[2]). *Then*

$$(\overset{*}{*}) \qquad \sum_{j\in J} f_j g_j \equiv 0 \ (\text{with } f_j\in\mathscr{S}_\omega,\ g_j\in\mathscr{D},\ J \text{ finite}) \Rightarrow \sum_{j\in J} f_j g_j = 0.$$

Proof: It is enough to show that $\sum_{j\in J} f_j g_j$ can be written in the form $\sum_{h\in H}\varphi_{A_h}g_h$, where $\varphi_{A_h}\in\mathscr{A}$, $\omega(A_h)=A_h$, $A_{h'}\cap A_{h''}=\emptyset$ if $h'\neq h''$, $g_h\in\mathscr{D}$ and H finite. In fact from $\sum_{h\in H}\varphi_{A_h}g_h\equiv 0$ we deduce:

$$\varphi_{A_h}g_h \equiv 0 \quad \text{for each} \quad h\in H.$$

Hence by (*)

$$\varphi_{A_h}g_h = 0 \quad \text{for each} \quad h\in H,$$

that is,

$$\sum_{h\in H}\varphi_{A_h}g_h = 0,$$

and the proposition is proved.

Therefore it remains to show that $\sum_{j\in J} f_j g_j$ can be represented in the desired form. If for each j,

$$f_j = \sum_{i\in I(j)} \lambda_i^{(j)}\varphi_{A_{i,j}}$$

then

$$\sum_{j\in J} f_j g_j = \sum_{j\in J}\Big(\sum_{i\in I(j)}\lambda_i^{(j)}\varphi_{A_{i,j}}\Big)g_j = \sum_{j\in J}\Big(\sum_{i\in I(j)}\varphi_{A_{i,j}}\,g_j^{(i)}\Big)$$

where $g_j^{(i)}=\lambda_i^{(j)}g_j(\in\mathscr{D})$. Hence

$$\sum_{j\in J} f_j g_j = \sum_{(j,i)\in\bigcup_{j\in J}\{j\}\times I(j)}\varphi_{A_{i,j}}\,g_j^{(i)} = \sum_{h\in H}\varphi_{A_h}g_h$$

---

[1]) If $\varphi_A\in\mathscr{A}$, we define $\omega(A)$ by $\varphi_{\omega(A)}=\omega(\varphi_A)$.

[2]) Note that $\mathscr{S}_\omega$ is dense in $\omega(\mathscr{A})$.

where $\varphi_{A_h} \in \mathscr{A}, \omega(A_h) = A_h$, $g_h \in \mathscr{D}$ and H is finite. A standard device shows that the sets A_h can be actually chosen disjoint. Thus the proof is complete.

Proposition 8. – *Let X be a locally compact space, μ a positive Radon measure on X with* Supp $\mu = X$. *Let $\mathscr{A} \subset M_R^\infty(X,\mu)$ be an algebra having the property $(\mathscr{E})$ and closed under pointwise convergence of bounded sequences and suppose that ω is a lifting of $\mathscr{A}$. Let $\mathscr{D} \subset C_R^b(X)$, $\mathscr{D} \ni 1$ be a closed subalgebra of $C_R^b(X)$ containing a countable dense set. Assume that:*

$$(\star) \qquad \varphi_A \in \mathscr{A}, \quad \omega(A) = A, \quad g \in \mathscr{D} \quad \textit{and} \quad \varphi_A g \equiv 0 \Rightarrow \varphi_A g = 0.$$

There is then a lifting ρ of $M_R^\infty(X,\mu)$ with the following properties:

8.1) $\rho|\mathscr{A} = \omega$;

8.2) $\rho(f) = f$ *if* $f \in \mathscr{D}$.

Proof: Since $\mathscr{A}$ has the property $(\mathscr{E})$, there is a lifting ρ' of $M_R^\infty(X,\mu)$ satisfying $\rho'|\mathscr{A} = \omega$. Let now $N \subset X$ be a $\mu^\bullet$-negligible set such that $\rho'(f)(t) = f(t)$ for all $f \in \mathscr{D}$ and $t \in \complement N$.

Let $\mathscr{F} = \{A | \varphi_A \in \mathscr{A}\}$; we recall that $\mathscr{F}$ is a tribe. As in Proposition 7 let $\mathscr{S}_\omega$ be the algebra consisting of all functions f of the form $f = \sum_{i=1}^{n} \lambda_i \varphi_{A_i}$, where $\lambda_i \in R$ and $A_i \in \mathscr{F}$, $\omega(A_i) = A_i$.

Let $\mathscr{X}$ be the algebra spanned by $\mathscr{S}_\omega$ and $\mathscr{D}$; $\mathscr{X}$ is the set of all functions of the form

$$\sum_{j \in J} f_j g_j$$

with $f_j \in \mathscr{S}_\omega$, $g_j \in \mathscr{D}$ and J finite.

Let $t \in X$. For $u \in \mathscr{X}$ define $\chi_t(u) = u(t)$; if $v \in \mathscr{X}$ and $v \equiv u$ then $v - u \equiv 0$ and by $(\star)$ of Proposition 7, $v - u = 0$. Thus $\chi_t(u) = \chi_t(v)$. Let now $\tilde{v} \in \tilde{\mathscr{X}} = \{\tilde{u} | u \in \mathscr{X}\}$; define

$$\chi_t(\tilde{v}) = u(t)$$

if $u \in \mathscr{X}$ and $\tilde{u} = \tilde{v}$. Then χ_t is well defined and is a character[1]) of $\tilde{\mathscr{X}}$. Let $\tilde{\mathscr{X}}_1$ be the closure of $\tilde{\mathscr{X}}$ in $L_R^\infty(X,\mu)$; χ_t can be extended by continuity to $\tilde{\mathscr{X}}_1$ and of course remains a character. Denote again by χ_t a (character) extension of χ_t to $L_R^\infty(X,\mu)$ (see Proposition 1).

---

[1]) Note that if $u \in \mathscr{X}$ then $|u| \in \mathscr{X}$ (represent u in the form $\sum_h \varphi_{A_h} g_h$ with disjoint A_h, and use the fact that $g \in \mathscr{D}$ implies $|g| \in \mathscr{D}$ since $\mathscr{D}$ is a closed algebra). Let now $\tilde{v} \in \tilde{\mathscr{X}}$, $\tilde{v} \geqslant 0$. There is then $u \in \mathscr{X}$ such that $\tilde{v} = \tilde{u}$ and $u(x) \geqslant 0$ $\mu^\bullet$-almost everywhere; it follows that $u \equiv |u|$ and hence $\chi_t(\tilde{v}) \geqslant 0$. We deduce that χ_t is continuous on $\tilde{\mathscr{X}}$.

Let now $f \in \mathscr{A}$. Since $\mathscr{S}_\omega$ is dense in $\omega(\mathscr{A})$, there is a sequence (f_n) of functions in $\mathscr{S}_\omega$ such that

$$\lim_n \|\omega(f)-f_n\|_\infty = 0;$$

in particular we deduce for each $t \in N$,

$$\omega(f)(t) = \lim_n f_n(t) = \lim_n \chi_t(\tilde{f}_n) = \chi_t(\tilde{f}).$$

We may now define ρ on $M_R^\infty(X,\mu)$ as follows:

$$\rho(f)(t) = \begin{cases} \chi_t(\tilde{f}) & \text{if } t \in N, \\ \rho'(f)(t) & \text{if } t \in \mathsf{C}N. \end{cases}$$

It is easily verified that ρ is a lifting of $M_R^\infty(X,\mu)$ satisfying 8.1) and 8.2).

Theorem 4. – *Let $X = S \times T$ where S and T are compact spaces and let μ be a positive Radon measure on X with* $\operatorname{Supp}\mu = X$. *Suppose that:*

4.1) T is metrizable;

4.2) ρ_1 is a strong lifting of $M_R^\infty(S, pr_S(\mu))$;

4.3) If $B \subset S$ is $pr_S(\mu)$-measurable, $U \subset T$ is open and $\mu(B \times U) = 0$ then $pr_S(\mu)(B) \cdot pr_T(\mu)(U) = 0$.

Then there is a strong lifting ρ of $M_R^\infty(X,\mu)$ such that

$$\rho(f \otimes g) = \rho_1(f) \otimes g$$

for all $f \in M_R^\infty(S, pr_S(\mu))$ and all $g \in C_R(T)$.

Proof: Let $\mathscr{A}$ be the algebra

$$\{f \circ pr_S + g \mid f \in M_R^\infty(S, pr_S(\mu)), \quad g \in \mathscr{N}^\infty\};$$

by Proposition 5, $\mathscr{A}$ is an admissible algebra. By Proposition 6, if we write $\omega(f \circ pr_S) = \rho_1(f) \circ pr_S$ for $f \in M_R^\infty(S, pr_S(\mu))$, then ω is a lifting of $\mathscr{A}$.

If we define

$$\mathscr{D} = \{f \circ pr_T \mid f \in C_R(T)\},$$

then $\mathscr{D} \subset C_R(X)$, $\mathscr{D} \ni 1$ and $\mathscr{D}$ is a closed subalgebra of $C_R(X)$ containing a countable dense set.

We shall show that $\mathscr{A}$ and $\mathscr{D}$ satisfy condition (*) of Proposition 8:

Let $\varphi_A \in \mathscr{A}$, $\omega(A) = A$ and $g \in \mathscr{D}$ be such that $\varphi_A g \equiv 0$. Then $A = B \times T$ with $\rho_1(B) = B$. Let $G = \{x \in X \mid g(x) \neq 0\}$; then $G = S \times U$ with $U \subset T$ open. For $x = (s,t) \in S \times T$ we have

$$(\varphi_A g)(s,t) \neq 0 \Leftrightarrow (\varphi_A \varphi_G)(s,t) \neq 0 \Leftrightarrow (s,t) \in A \cap G = B \times U.$$

We deduce that

$$\varphi_A g \equiv 0 \Rightarrow \mu(B \times U) = 0 \Rightarrow pr_S(\mu)(B) \cdot pr_T(\mu)(U) = 0;$$

since $B=\rho_1(B)$ and U is open this implies that either $B=\emptyset$ or $U=\emptyset$, that is, $B\times U=\emptyset$. Thus $\varphi_A g=0$.

By Proposition 8, there is a lifting ρ of $M_R^\infty(X,\mu)$ such that

$$\rho(f\otimes 1)=\rho_1(f)\otimes 1 \quad \text{for all} \quad f\in M_R^\infty(S,pr_S(\mu))$$

and

$$\rho(1\otimes g)=1\otimes g \quad \text{for all} \quad g\in C_R(T),$$

whence

$$\rho(f\otimes g)=\rho_1(f)\otimes g \quad \text{for all} \quad f\in M_R^\infty(S,pr_S(\mu)) \quad \text{and} \quad g\in C_R(T).$$

The theorem is therefore proved.

Note that 4.3) is satisfied if μ is a product measure.

Let $(X_i)_{i\in J}$ be a family of compact spaces and let

$$X=\prod_{i\in J} X_i.$$

For every $I\subset J$ let

$$X_I=\prod_{i\in I} X_i$$

and let pr_I be the projection mapping of X onto X_I. If $I'\subset I$ we denote by $pr_{I',I}$ the projection mapping of X_I onto $X_{I'}$.

In the next proposition we collect some results that will be needed later:

Proposition 9. – *Let $I\subset J$ and let λ be a positive Radon measure on X_I. Then:*

9.1) If $f\in C_R(X_I)$, then there is a countable set $I_0\subset I$ and $f_0\in C_R(X_{I_0})$ such that $f=f_0\circ pr_{I_0,I}$.

9.2) If $f\in M_R^\infty(X_I,\lambda)$, then there is a countable set $I_1\subset I$ and a function $f_1\in M_R^\infty(X_{I_1},pr_{I_1,I}(\lambda))$ such that $f\equiv f_1\circ pr_{I_1,I}$.

Proof: 9.1) Let $f\in C_R(X_I)$. There is then a sequence $(I(n))$ of finite parts of I and a sequence (f_n) of functions such that:

1) $f_n\in C_R(X_{I(n)})$ for each n;

2) $\lim_n \|f_n\circ pr_{I(n),I}-f\|_\infty=0$.

Let now $I_0=\bigcup_n I(n)$. Then I_0 is countable. Let $f_n'=f_n\circ pr_{I(n),I_0}$ for every n. Clearly $f_n'\in C_R(X_{I_0})$ and $f_n'\circ pr_{I_0,I}=f_n\circ pr_{I(n),I}$, whence $(f_n'\circ pr_{I_0,I})$ converges to f. But (f_n') is a Cauchy sequence in $C_R(X_{I_0})$ since

$$\|f_n'-f_m'\|_\infty=\|f_n'\circ pr_{I_0,I}-f_m'\circ pr_{I_0,I}\|_\infty.$$

Hence there is $f_0\in C_R(X_{I_0})$ such that (f_n') converges to f_0. It follows that $f_0\circ pr_{I_0,I}=f$.

9.2) Let $f \in M_R^\infty(X_I, \lambda)$. There is then a countable set $I_1 \subset I$ and a sequence (f_n) of functions in $C_R(X_{I_1})$ such that:

3) $\sup_n \|f_n\|_\infty < \infty$;

4) $(f_n \circ pr_{I_1,I})$ converges to f almost everywhere.

Let now $f_1 = \limsup_n f_n$. Then

$$f_1 \circ pr_{I_1,I} = \limsup_n f_n \circ pr_{I_1,I}$$

and from 4) it follows that

$$f_1 \circ pr_{I_1,I} \equiv f$$

and hence the proposition is proved

Let now μ be *a positive Radon measure on* $X\,(=X_J)$ *with* $\operatorname{Supp}\mu = X$. *For every* $I \subset J$ *let* $\mu_I = pr_I(\mu)$. *Let* $\mathscr{I}$ *be the set of all pairs* (I, ρ^I) *where* $I \subset J$ *and* ρ^I *is a strong lifting of* $M_R^\infty(X_I, \mu_I)$.

We now *order* the set $\mathscr{I}$ as follows:

We write $(I', \rho^{I'}) \leqslant (I'', \rho^{I''})$ if $I' \subset I''$ and $\rho^{I''}(f \circ pr_{I',I''}) = \rho^{I'}(f) \circ pr_{I',I''}$ for all $f \in M_R^\infty(X_{I'}, \mu_{I'})$. It is easily seen that this is an order relation on $\mathscr{I}$.

Proposition 10. – *The set* $\mathscr{I}$ *is inductive for the above order relation.*

Proof: Let $\Phi = (I(j), \rho^{I(j)})_{j \in H}$ be a totally ordered family of elements of $\mathscr{I}$ (we suppose that $j' \leqslant j''$ if and only if $(I(j'), \rho^{I(j')}) \leqslant (I(j''), \rho^{I(j'')})$). Let $I = \bigcup_{j \in H} I(j)$. There are two possibilities:

A) *There is no countable cofinal part in* H.

Let $f \in M_R^\infty(X_I, \mu_I)$. There is then a countable set $I_0 \subset I$ and $f_0 \in M_R^\infty(X_{I_0}, \mu_{I_0})$ such that $f \equiv f_0 \circ pr_{I_0,I}$. Clearly there is $j_1 \in H$ such that $I_0 \subset I(j_1)$. If we set $f_1 = f_0 \circ pr_{I_0, I(j_1)}$ we have

$$f_1 \in M_R^\infty(X_{I(j_1)}, \mu_{I(j_1)})$$

and

$$f_1 \circ pr_{I(j_1),I} = (f_0 \circ pr_{I_0, I(j_1)}) \circ pr_{I(j_1),I} = f_0 \circ pr_{I_0,I} \equiv f.$$

Define now

$$\rho(f) = \rho^{I(j_1)}(f_1) \circ pr_{I(j_1),I}.$$

If $f \equiv f_2 \circ pr_{I(j_2),I}$ with $f_2 \in M_R^\infty(X_{I(j_2)}, \mu_{I(j_2)})$ and $j_2 \neq j_1$ then either $j_1 < j_2$ or $j_2 < j_1$. Suppose for instance that $j_1 < j_2$. Note that

$$f_2 \circ pr_{I(j_2),I} \equiv f_1 \circ pr_{I(j_1),I}$$

implies

$$f_2 \circ pr_{I(j_2),I} \equiv (f_1 \circ pr_{I(j_1),I(j_2)}) \circ pr_{I(j_2),I},$$

that is

$$f_2 \equiv f_1 \circ pr_{I(j_1),I(j_2)}.$$

We deduce

$$\rho^{I(j_2)}(f_2) = \rho^{I(j_2)}(f_1 \circ pr_{I(j_1),I(j_2)}) = \rho^{I(j_1)}(f_1) \circ pr_{I(j_1),I(j_2)}$$

and therefore

$$\rho^{I(j_2)}(f_2) \circ pr_{I(j_2),I} = \rho^{I(j_1)}(f_1) \circ pr_{I(j_1),I}.$$

Thus $\rho(f)$ is well defined. It is easily seen that ρ is a lifting of $M_R^\infty(X_I, \mu_I)$ and that ρ is strong (approximate every $f \in C_R(X_I)$ by functions depending only on a finite number of coordinates). It is also easily checked that (I, ρ) is a majorant for Φ.

B) *There is a countable cofinal part in H.*
We may assume that this is the set of elements of an increasing sequence $(j(n))$. Let $\mathcal{N}_I^\infty$ be the ideal of $M_R^\infty(X_I, \mu_I)$ consisting of all $\mu_I^\bullet$-negligible functions. For each n define

$$\mathscr{A}_n = \{f \circ pr_{I(j_n),I} + g \mid f \in M_R^\infty(X_{I(j_n)}, \mu_{I(j_n)}), g \in \mathcal{N}_I^\infty\};$$

then $\mathscr{A}_n$ is an admissible subalgebra of $M_R^\infty(X_I, \mu_I)$ (see Proposition 5). We denote by ω_n the lifting of $\mathscr{A}_n$ corresponding to $\rho^{I(j_n)}$, defined as in Proposition 6. Recall that if $u \in \mathscr{A}_n$ and if $u \equiv f \circ pr_{I(j_n),I}$ for some $f \in M_R^\infty(X_{I(j_n)}, \mu_{I(j_n)})$ then

$$\omega_n(u) = \rho^{I(j_n)}(f) \circ pr_{I(j_n),I}.$$

It is easily seen that $\omega_n | \mathscr{A}_m = \omega_m$ if $m \leqslant n$ and that the admissible subalgebra of $M_R^\infty(X_I, \mu_I)$ spanned by $\bigcup_n \mathscr{A}_n$ is precisely $M_R^\infty(X_I, \mu_I)$. By Theorem 2, section 1, chapter 4, there is a lifting ω of $M_R^\infty(X_I, \mu_I)$ such that $\omega | \mathscr{A}_n = \omega_n$ for all n. It is obvious that the lifting ω of $M_R^\infty(X_I, \mu_I)$ is strong. We shall verify that for each n, $(I, \omega) \geqslant (I(j_n), \rho^{I(j_n)})$.

Let $h \in M_R^\infty(X_{I(j_n)}, \mu_{I(j_n)})$. Then $h \circ pr_{I(j_n),I} \in \mathscr{A}_n$ and we have:

$$\omega(h \circ pr_{I(j_n),I}) = \omega_n(h \circ pr_{I(j_n),I}) = \rho^{I(j_n)}(h) \circ pr_{I(j_n),I}.$$

Thus $(I, \omega) \geqslant (I(j_n), \rho^{I(j_n)})$ for every n and hence (I, ω) is a majorant for Φ. This completes the proof of Proposition 10.

With the notation introduced above we may now state:

Theorem 5. – *Let* $X = \prod_{i \in J} X_i$, *where* $(X_i)_{i \in J}$ *is a family of compact spaces and let* μ *be a positive Radon measure on* X *with* $\operatorname{Supp} \mu = X$. *Suppose that:*

5.1) X_i *is metrizable for each* $i \in J$.

5.2) *If* $I \subset J$, $i \in J - I$, $B \subset X_I$ *is* $\mu_I^\bullet$*-measurable,* $U \subset X_i$ *is open and* $\mu_{I \cup \{i\}}(B \times U) = 0$, *then* $\mu_I(B) \cdot \mu_{\{i\}}(U) = 0$.

Let

$$\mathscr{I} = \{(I,\rho^I) | I \subset J \text{ and } \rho^I \text{ is a strong lifting of } M_R^\infty(X_I,\mu_I)\}.$$

If (I,ρ^I) *is maximal in* $\mathscr{I}$, *we have* $I=J$ *and hence* ρ^I *is a strong lifting of* $M_R^\infty(X,\mu)$.

Proof: Let $(I,\rho^I)\in\mathscr{I}$ be maximal in $\mathscr{I}$ and suppose $I\neq J$. Let $i\in J-I$. By Theorem 4, there is a strong lifting ρ^* of $M_R^\infty(X_{I\cup\{i\}},\mu_{I\cup\{i\}})$ such that

$$\rho^*(f\otimes g) = \rho^I(f)\otimes g$$

for all $f\in M_R^\infty(X_I,\mu_I)$ and all $g\in C_R(X_i)$. For $f\in M_R^\infty(X_I,\mu_I)$ we have (below 1 is the constant function on X_i):

$$\rho^*(f\circ pr_{I,I\cup\{i\}}) = \rho^*(f\otimes 1) = \rho^I(f)\otimes 1 = \rho^I(f)\circ pr_{I,I\cup\{i\}}.$$

Thus $(I\cup\{i\},\rho^*)$ is a strict majorant of (I,ρ^I); this yields a contradiction and hence Theorem 5 is proved.

Corollary. – *If the conditions of Theorem 5 are satisfied then* (X,μ) *has the strong lifting property.*

We shall now give two examples of couples (X,μ) satisfying the conditions of Theorem 5.

Example 1. – $X=\prod_{i\in J} X_i$, where $(X_i)_{i\in J}$ is a family of metrizable compact spaces, and $\mu = \bigotimes_{i\in J}\mu_i$, where for each $i\in J$, μ_i is a positive Radon measure on X_i with $\operatorname{Supp}\mu_i = X_i$.

Example 2. – Let J be a set and $X=\prod_{i\in J} X_i$, where $X_i=\{0,1\}$ for all $i\in J$. For each $p\in[0,1]$ let λ_p be the measure on $\{0,1\}$ defined by $\lambda_p(\{0\})=p$ and $\lambda_p(\{1\})=1-p$. For each $i\in J$ let $\lambda_p^i=\lambda_p$. For every $I\subset J$ define the measure λ_p^I on X_I by:

$$\lambda_p^I = \bigotimes_{i\in I}\lambda_p^i.$$

Clearly λ_p^J is a measure on $X=X_J$ of total mass one and if $0<p<1$, then $\operatorname{Supp}\lambda_p^J = X$ (note that the measure of every non-void "cylindric set" is >0). Note also that for each $I\subset J$

$$pr_I(\lambda_p^J) = \lambda_p^I.$$

Now the mapping $p\to\lambda_p^J$ of $[0,1]$ into $\mathscr{M}_R(X)$ is vaguely continuous[1]) (use the fact that every $f\in C_R(X)$ can be approximated by

---

[1]) That is, $p\to\lambda_p^J$ is continuous as a mapping of $[0,1]$ into $\mathscr{M}_R(X)$ endowed with the topology $\sigma(\mathscr{M}_R(X), C_R(X))$.

functions depending only on a finite number of coordinates). Let now β be a positive Radon measure on $[0,1]$ of total mass one and such that $\beta(\{1\})=\beta(\{0\})=0$. Define the measure μ on X by

$$\mu = \int_{[0,1]} \lambda_p^J d\beta(p)$$

that is,

$$\mu(f) = \int_{[0,1]} \lambda_p^J(f) d\beta(p), \quad \text{for all} \quad f \in C_R(X).$$

It is clear that $\operatorname{Supp}\mu = X$ and that 5.1) of Theorem 5 is satisfied. We shall show that the couple (X,μ) satisfies also condition 5.2).

We recall that if $E \subset X$ is $\mu^\bullet$-measurable, then the set H of all $p \in [0,1]$ for which E is not $(\lambda_p^J)^\bullet$-measurable is $\beta^\bullet$-negligible, the function $p \to \lambda_p^J(E)$ defined for $p \notin H$ is $\beta^\bullet$-measurable and we have

$$\mu(E) = \int_{[0,1]} \lambda_p^J(E) d\beta(p)$$

(see [15], chapter 5, p. 19 and pp. 25–26).

Note also that for each $I \subset J$,

$$\mu_I = pr_I(\mu) = \int_{[0,1]} pr_I(\lambda_p^J) d\beta(p) = \int_{[0,1]} \lambda_p^I d\beta(p).$$

Let now $I \subset J$, $i \in J - I$; suppose that $B \subset X_I$ is $\mu_I^\bullet$-measurable, $U \subset X_i$ is open and $\mu_{I \cup \{i\}}(B \times U) = 0$. We deduce

$$\int_{[0,1]} \lambda_p^I(B) \lambda_p(U) d\beta(p) = 0,$$

whence $\lambda_p^I(B)\lambda_p(U)=0$ for almost every $p \in [0,1]$ with respect to β. If $U=\emptyset$ then $\mu_{\{i\}}(U)=0$ and hence $\mu_I(B)\cdot\mu_{\{i\}}(U)=0$. If $U \neq \emptyset$, then $\lambda_p(U) \neq 0$ for all $p \in (0,1)$ and hence for almost every $p \in [0,1]$ with respect to β. It follows that $\lambda_p^I(B)=0$ for almost every $p \in [0,1]$ with respect to β, whence $\mu_I(B)=0$; again we deduce $\mu_I(B)\cdot\mu_{\{i\}}(U)=0$. This completes the proof.

Remark. – It should be noted that the measure μ constructed in Example 2 is not necessarily equivalent with a "product measure".

In fact, take for instance $J=R$ and $\beta = \frac{1}{2}\varepsilon_p + \frac{1}{2}\varepsilon_q$ with $p,q \in (0,1)$ and $p \neq q$. Then (with the notations of Example 2),

$$\mu = \tfrac{1}{2}\lambda_p^J + \tfrac{1}{2}\lambda_q^J.$$

Note that if $I \subset R$ is an infinite set, then the measures λ_p^I and λ_q^I defined on X_I are mutually singular.

Let now $I_1=(0,+\infty)$ and $I_2=(-\infty,0]$. There are then sets[1]) $E_p^1 \subset X_{I_1}$, $E_q^1 \subset X_{I_1}$ mutually disjoint and sets $E_p^2 \subset X_{I_2}$, $E_q^2 \subset X_{I_2}$ mutually disjoint such that:

$$\lambda_p^{I_1}(E_p^1)=1,\ \lambda_q^{I_1}(E_p^1)=0, \qquad \lambda_p^{I_2}(E_p^2)=1,\ \lambda_q^{I_2}(E_p^2)=0,$$
$$\lambda_p^{I_1}(E_q^1)=0,\ \lambda_q^{I_1}(E_q^1)=1, \qquad \lambda_p^{I_2}(E_q^2)=0,\ \lambda_q^{I_2}(E_q^2)=1.$$

Let $A=E_p^1\times E_q^2$. This is a set in $X=X_{I_1}\times X_{I_2}$ which is measurable with respect to λ_p^J and λ_q^J. We have:

$$\lambda_p^J(A)=\lambda_p^J(E_p^1\times E_q^2)=\lambda_p^{I_1}(E_p^1)\cdot\lambda_p^{I_2}(E_q^2)=0$$

and

$$\lambda_q^J(A)=\lambda_q^J(E_p^1\times E_q^2)=\lambda_q^{I_1}(E_p^1)\cdot\lambda_q^{I_2}(E_q^2)=0.$$

Hence

$$\mu(A)=\tfrac{1}{2}\,\lambda_p^J(A)+\tfrac{1}{2}\,\lambda_q^J(A)=0.$$

Assume that μ is equivalent with a product measure ν. Then, for each $I\subset R$, $\mu_I=pr_I(\mu)$ and $\nu_I=pr_I(\nu)$ are equivalent measures on X_I. We have:

$$\mu_{I_1}(E_p^1)>0,\quad \mu_{I_2}(E_q^2)>0 \Rightarrow \nu_{I_1}(E_p^1)>0,\quad \nu_{I_2}(E_q^2)>0$$

whence

$$\nu(A)=\nu(E_p^1\times E_q^2)=\nu_{I_1}(E_p^1)\cdot\nu_{I_2}(E_q^2)>0.$$

Since this is a contradiction, we have proved that the measure μ defined above *is not equivalent with a product measure.*

We shall close this section with several indications concerning *one more example.*

Let X be a locally compact space and $\mu\neq 0$ a positive Radon measure on X. Denote by $\mathscr{H}(X,\mu)$ the group of all bijections $s\colon X\to X$ having the following two properties:

(1) $s(A)$ and $s^{-1}(A)$ are $\mu^\bullet$-measurable if A is;

(2) $s(A)$ and $s^{-1}(A)$ are $\mu^\bullet$-negligible if A is. We remark that for each $s\in\mathscr{H}(X,\mu)$ the mapping $f\to f\circ s$ is an isomorphism of $M_R^\infty(X,\mu)$ onto itself; clearly $f\equiv g$ implies $f\circ s\equiv f\circ g$.

Let now $\mathscr{J}\subset\mathscr{H}(X,\mu)$. A *linear lifting* or a *lifting* ρ of $M_R^\infty(X,\mu)$ *is said to commute with* $\mathscr{J}$ *if*

$$\rho(f\circ s)=\rho(f)\circ s$$

for all $f\in M_R^\infty(X,\mu)$ *and* $s\in\mathscr{J}$.

Suppose now that X is a *locally compact group and* μ *a left invariant Haar measure on* X. *Let* $\mathscr{J}$ *be the group of left translations of* X. *If* ρ

[1]) We may and will assume that the sets considered below are Borel sets.

is a linear lifting of $M_R^\infty(X,\mu)$ *which commutes with* $\mathscr{J}$, *then* ρ *is strong.* In fact, for each $s \in X$ let $\gamma(s): \mathscr{K}_R(X) \to \mathscr{K}_R(X)$ be defined by

$$(\gamma(s)f)(x) = f(s^{-1}x)$$

(for $f \in \mathscr{K}_R(X)$ and $x \in X$). Let now $f \in \mathscr{K}_R(X)$ be fixed. The mappings $s \to s^{-1}$ of X into X and $t \to \gamma(t)f$ of X into $M_R^\infty(X,\mu)$ (endowed with the supremum norm $f \to \|f\|_\infty$) are continuous. Since

$$\|\rho(f)\|_\infty \leqslant \|f\|_\infty$$

for all $f \in M_R^\infty(X,\mu)$, we deduce that $s \to \rho(\gamma(s^{-1})f)$ is a continuous mapping of X into $M_R^\infty(X,\mu)$; in particular[1]) $s \to \rho(\gamma(s^{-1})f)(e) = \rho(f)(s)$ is continuous. Since f and $\rho(f)$ are continuous and concide $\mu^\bullet$-almost everywhere, we deduce $\rho(f) = f$. Since $f \in \mathscr{K}_R(X)$ was arbitrary, we conclude that ρ is strong.

If ρ is a lifting, this last assertion follows from Theorem 1. If ρ is only a linear lifting we reason as follows: Let $g \in C_R^b(X)$, $g \geqslant 0$ and let $\mathscr{F}_g = \{f \in \mathscr{K}_R(X) | 0 \leqslant f \leqslant g\}$. We have $g = \sup \mathscr{F}_g$. But $\rho(g) \geqslant \rho(f) = f$ for each $f \in \mathscr{F}_g$, whence $\rho(g) \geqslant \sup \mathscr{F}_g = g$. Let now $g \in C_R^b(X)$ arbitrary and let λ be a constant such that $\lambda + g \geqslant 0$ and $\lambda - g \geqslant 0$. We deduce $\lambda + \rho(g) = \rho(\lambda + g) \geqslant \lambda + g$ and $\lambda - \rho(g) = \rho(\lambda - g) \geqslant \lambda - g$; hence $\rho(g) = g$. Since $g \in C_R^b(X)$ was arbitrary we conclude that ρ is strong.

Using a delicate analysis (and approximation with Lie groups) we showed in [72] that:

(LCG) *If* X *is a locally compact group and* μ *a left invariant Haar measure on* X, *then there exists a lifting of* $M_R^\infty(X,\mu)$ *commuting with the left translations of* X.

As we remarked above, a lifting commuting with the left translations of X is necessarily *strong*. Hence:

If X *is a locally compact group and* μ *a left invariant Haar measure on* X *then* (X,μ) *has the strong lifting property.*

For the proof of (LCG) see the authors' paper [72]. In connection with liftings commuting with certain mappings see also [54] and [60].

3. An example and several related results

Let R be the real line and β the Lebesgue measure on R. Denote by $C_R^b(R,+)$ the algebra of all bounded real-valued functions on R, *continuous on the right*, and by $C_R^b(R,-)$ the algebra of all bounded real-valued functions on R, *continuous on the left*. With this notation we may state and prove the following:

[1]) We denote by e the unit element of the *group* X.

Theorem 6. – *There is a lifting ρ of $M_R^\infty(R,\beta)$ such that $\rho(f)=f$ for every $f\in C_R^b(R,+)$ (respectively, for every $f\in C_R^b(R,-)$).*

Proof: We shall consider only the case of $C_R^b(R,+)$; the case of $C_R^b(R,-)$ can be treated similarly.

For each $x\in R$ and $n\in N^*=\{1,2,3,\dots\}$ let

$$I_n(x)=[x,x+1/n].$$

For $f\in M_R^\infty(R,\beta)$ and $n\in N^*$ define the function $f_n\colon R\to R$ by the equations

$$f_n(x)=\frac{1}{\beta(I_n(x))}\int_{I_n(x)} f(t)\,d\beta(t)=n\int_{I_n(x)} f(t)\,d\beta(t),\quad \text{for}\quad x\in R.$$

It is obvious that $f_n\in M_R^\infty(R,\beta)$ (in fact $f_n\in C_R^b(R)$) and that

$$\sup\{|f_n(x)|\,|n\in N^*,x\in X\}\leqslant N_\infty(f).$$

Let $\mathscr{U}$ be an ultrafilter on N^* finer than the Fréchet filter on N^*. Then, for each $x\in R$, $n\to f_n(x)$ has a limit $f_\infty(x)$ with respect to $\mathscr{U}$. Since the sequence $(f_n(x))$ converges almost everywhere to $f_\infty(x)$ (see for instance [121], p. 132) and since f_∞ is bounded on R, it follows that $f_\infty\in M_R^\infty(R,\beta)$. Define

$$\gamma(f)=f_\infty,\quad \text{for}\quad f\in M_R^\infty(R,\beta).$$

Then $\gamma\colon f\to\gamma(f)$ is a linear lifting of $M_R^\infty(R,\beta)$ and $\gamma(f)=f$ if $f\in C_R^b(R,+)$. Let θ and θ' be the lower and upper densities associated with γ and let $\mathscr{G}(\gamma)$ be the set of all linear liftings σ of $M_R^\infty(R,\beta)$ satisfying the inequalities

$$\varphi_{\theta(A)}\leqslant\sigma(\varphi_A)\leqslant\varphi_{\theta'(A)}$$

for every $\beta^\bullet$-measurable set $A\subset R$ (see Definition 6, section 2, chapter 3). Then there is a lifting ρ of $M_R^\infty(R,\beta)$ belonging to the set $\mathscr{G}(\gamma)$ (see Theorem 2, section 2, chapter 3). It is clear that if $A=[a,b)$ where $a\in R$, $b\in R$, $a<b$, then $\rho(\varphi_A)=\gamma(\varphi_A)=\varphi_A$ (since in this case $\varphi_A\in C_R^b(R,+)$).

We shall now show that $\rho(g)\geqslant g$ for each $g\in C_R^b(R,+)$, $g\geqslant 0$. In fact, let Φ_g be the set of all functions $\alpha\varphi_{[a,b)}\leqslant g$ with $\alpha\geqslant 0$, $a\in R$, $b\in R$, $a<b$. It is easy to see that $\sup\Phi_g=g$ (pointwise supremum). We deduce $\rho(g)\geqslant\rho(u)=u$ for each $u\in\Phi_g$, whence $\rho(g)\geqslant g$.

Let now $f\in C_R^b(R,+)$ and let c be a constant such that $c+f\geqslant 0$, $c-f\geqslant 0$. Since $\rho(c)=c$ we deduce

$$c+\rho(f)=\rho(c+f)\geqslant c+f$$

and

$$c-\rho(f)=\rho(c-f)\geqslant c-f.$$

Hence $\rho(f)=f$ and the theorem is completely proved.

Remark. – From Theorem 6 above (use also Theorem 3, section 3, chapter 3) we deduce in particular that if

$$\mathscr{H} \subset C^b_{\mathbf{R}}(R,+) \quad \text{(respectively, } \mathscr{H} \subset C^b_{\mathbf{R}}(R,-)\text{)}$$

is a set of positive functions directed for the relation $\leqslant$, then $f_\infty = \sup \mathscr{H}$ (pointwise supremum) is $\beta^\bullet$-measurable and

$$\int_R^\bullet f_\infty \, d\beta = \sup_{f \in \mathscr{H}} \int_R^\bullet f \, d\beta.$$

In connection with this remark see also Lemma 4.1, p. 112, [121].

An example. – For each $x \in R$ and $y > 0$ let

$$U_y(x) = [x, x+y) \cup (-x-y, -x).$$

Then there is a topology Σ on R such that $\{U_y(x) | y > 0\}$ is a fundamental system of neighborhoods of x for each $x \in R$. If we denote by X *the interval* $[-1,1)$ *endowed with the topology induced by* Σ, *then:*

1) X is compact;

2) Every point $x \in X$ admits a countable fundamental system of neighborhoods;

3) There is a countable set dense in X;

4) X is not metrizable

(see [14], Topologie, chapter 9 (1958), p. 49). It is also easily seen that a function $f: X \to R$ belongs to $C_{\mathbf{R}}(X)$ if and only if

$$f(x) = f(x+) = f((-x)-) \quad \text{for each} \quad x \in X.$$

Below we shall denote by B *and* T *the intervals* $[0,1)$ *and* $[-1,1)$ respectively, when *endowed with the usual topological structure.* Let β be the Lebesgue measure on R and let β_B be the measure induced by β on B, i. e. the Lebesgue measure on $B = [0,1)$. Since every function on $[-1,1)$ continuous on the right is Borel and hence Lebesgue measurable, we may define the Radon measure μ on X by the equations:

$$\mu(f) = \int_B (f|B) \, d\beta_B \quad \text{for} \quad f \in C_{\mathbf{R}}(X).$$

The statement of Proposition 11 below and of the Corollary following it are taken from [15], chapter 6, p. 99.

Proposition 11. – 11.1) *We have* $\operatorname{Supp} \mu = X$.

11.2) *Let* $A \subset X$. *Then the following assertions are equivalent:*

i) $\mu^\bullet(A) = 0$.

ii) $\beta^\bullet(A) = 0$.

Proof: 11.1) Let $x \in X$, let $V \subset X$ be an open neighborhood of x and let $f \in C_R(X)$, $f \geqslant 0$ with $f(x) \neq 0$ and $\operatorname{Supp} f \subset V$. If $x \in [0,1)$, then $(f|B)(x) = f(x) \neq 0$. If $x \in [-1,0)$, then $-x \in (0,1]$, $f((-x)-) = f(x) \neq 0$ and hence $f(y) \neq 0$ for some $y \in B$. In either case it is obvious that

$$\mu(f) = \int_B (f|B)\, d\beta_B \neq 0$$

and hence $x \in \operatorname{Supp} \mu$.

11.2) i)⇒ii). Let $\varepsilon > 0$ and let $U \subset X$ open (for Σ) such that $U \supset A$ and $\mu^\bullet(U) \leqslant \varepsilon$. Define

$$\mathscr{H} = \{f \in C_R(X) | 0 \leqslant f \leqslant \varphi_U\}.$$

Then $\mathscr{H}$ is directed for the relation $\leqslant$ and its pointwise supremum is φ_U. For each $f: X \to R$ denote by f' the function on R to R defined by $f'(x) = f(x)$ if $x \in X$ and $f'(x) = 0$ if $x \notin X$; note that $f' \in C_R^b(R,+)$ for each $f \in C_R(X)$. We deduce (use the Remark following Theorem 6 above):

$$\beta^\bullet(U) = \sup_{f \in \mathscr{H}} \int_R f'\, d\beta = \sup_{f \in \mathscr{H}} 2 \int_B (f|B)\, d\beta_B = 2 \sup_{f \in \mathscr{H}} \mu(f) = 2\mu^\bullet(U) \leqslant 2\varepsilon.$$

Since $\varepsilon > 0$ was arbitrary it follows that $\beta^\bullet(A) = 0$ and hence the implication i)⇒ii) is proved.

ii)⇒i). Let $\varepsilon > 0$ and let $U \subset T$ open (for the Euclidean topology $\mathscr{E}$ of $[-1,1)$) such that $U \supset A$ and $\beta^\bullet(U) \leqslant \varepsilon$. Let $(f_\iota)_{\iota \in I}$ be a directed (for $\leqslant$) family of functions belonging to $\mathscr{K}_R(T)$ such that $0 \leqslant f_\iota \leqslant \varphi_U$, $\operatorname{Supp} f_\iota \subset U$ and $\sup_{\iota \in I} f_\iota = \varphi_U$ (pointwise supremum); in particular then, for each $\iota \in I$, $f_\iota(1-) = 0$. For every $\iota \in I$ define $g_\iota: [-1,1) \to R$ by

$$g_\iota(x) = \begin{cases} f_\iota(x) + f_\iota(-x) & \text{if } x \in (-1,1), \\ f_\iota(-1) & \text{if } x = -1. \end{cases}$$

It is immediate thet g_ι is continuous on $[-1,1)$ (for $\mathscr{E}$), that $g_\iota(-x) = g_\iota(x)$ for each $x \in (-1,1)$ and that $g_\iota(-1) = g_\iota(1-)$. Hence $g_\iota \in C_R(X)$ for each $\iota \in I$. Now $(g_\iota)_{\iota \in I}$ is directed (for $\leqslant$) and if g is the pointwise supremum of $(g_\iota)_{\iota \in I}$, we deduce $g \geqslant \varphi_A$, whence

$$\sup_{\iota \in I} \mu(g_\iota) \geqslant \mu^\bullet(A).$$

But for each $\iota \in I$ we have:

$$\begin{aligned} \mu(g_\iota) &= \int_B (g_\iota|B)\, d\beta_B = \int_{[0,1)} f_\iota(x)\, d\beta_B(x) + \int_{[0,1)} f_\iota(-x)\, d\beta_B(x) \\ &= \int_{[-1,1)} f_\iota\, d\beta_T \leqslant \beta_T^\bullet(U) = \beta^\bullet(U) \leqslant \varepsilon. \end{aligned}$$

Since $\varepsilon > 0$ was arbitrary it follows that $\mu^\bullet(A) = 0$. Hence the implication ii)⇒i) is also proved. This completes the proof of the proposition.

Below we shall denote by α the induced measure β_T, i. e. the Lebesgue measure on $T = [-1,1)$. From Proposition 11 we then deduce the following:

Corollary. – 1) *A function* $f: [-1,1) \to R$ *is* $\mu^\bullet$*-negligible if and only if it is* $\alpha^\bullet$*-negligible.*

2) *If* $f: [-1,1) \to R$ *is* $\mu^\bullet$*-measurable, then* f *is* $\alpha^\bullet$*-measurable.*

3) *We have* $M_R^\infty(X,\mu) \subset M_R^\infty(T,\alpha)$ *and for* f *and* g *in* $M_R^\infty(X,\mu)$ *we have* $f \equiv g$ *in* $M_R^\infty(X,\mu)$ *if and only if* $f \equiv g$ *in* $M_R^\infty(T,\alpha)$.

4) *If* ρ *is a lifting of* $M_R^\infty(T,\alpha)$, *then* $\rho|M_R^\infty(X,\mu)$ *is a lifting of* $M_R^\infty(X,\mu)$.

Proof: 1) is obvious.

2) Let $f:[-1,1)\to R$ be $\mu^\bullet$-measurable. We may suppose that f is bounded. There is then a sequence (f_n) of functions belonging to $C_R(X)$ such that

$$\lim_n f_n(x) = f(x) \quad \text{for all} \quad x \notin A,$$

where $\mu^\bullet(A)=0$. Since $\alpha^\bullet(A)=\beta^\bullet(A)=0$ and since each f_n is also $\alpha^\bullet$-measurable, we deduce that f is $\alpha^\bullet$-measurable.

3) Follows immediately from 1) and 2) and 4) is a consequence of 3).

We may now state and prove the following:

Theorem 7. – 7.1) *Let* ρ *be a lifting of* $M_R^\infty(R,\beta)$ *such that* $\rho(f)=f$ *for every* $f \in C_R^b(R,+)$[1]. *Let* ρ' *be the lifting of* $M_R^\infty(T,\alpha)$ *defined by the equations*[2]

$$\rho'(f)(t) = \rho(f')(t) \quad \textit{for} \quad f \in M_R^\infty(T,\alpha) \quad \textit{and} \quad t \in T.$$

Then $\sigma = \rho'|M_R^\infty(X,\mu)$ *is a strong lifting of* $M_R^\infty(X,\mu)$.

7.2) *There is a lifting* δ *of* $M_R^\infty(X,\mu)$ *such that*

$$\bigcup_{f \in C_R(X)} \{x|\delta(f)(x) \neq f(x)\} \doteq X.$$

Proof: 7.1) It is clear that σ is a lifting of $M_R^\infty(X,\mu)$ (use statement 4) of the above Corollary). Let $f \in C_R(X)$. Then $f' \in C_R^b(R,+)$ and hence

$$\sigma(f) = \rho'(f) = \rho(f')|X = f'|X = f.$$

Thus σ is a strong lifting of $M_R^\infty(X,\mu)$.

7.2) Let $\gamma: f \to \gamma(f)$ be a lifting of $M_R^\infty(R,\beta)$ such that $\gamma(f)=f$ for every $f \in C_R^b(R,-)$. Let χ be a character of $L_R^\infty(X,\mu)$ such that $\chi(\tilde{f}) = f(\frac{1}{2})$ for all $f \in C_R(X)$. For $f \in M_R^\infty(X,\mu)$ define

$$\delta(f)(t) = \begin{cases} \gamma(f')(t) & \text{if} \quad t \neq 0 \quad \text{and} \quad t \in \gamma(X) \cap X = (-1,1), \\ \chi(\tilde{f}) & \text{if} \quad t = 0 \quad \text{or} \quad t = -1. \end{cases}$$

Then $\delta: f \to \delta(f)$ is a lifting of $M_R^\infty(X,\mu)$. If $f = \varphi_{[-a,a)}$, $0<a<1$, then $f \in C_R(X)$ and $\delta(f)(t) \neq f(t)$ for $t=-a$ or $t=a$. Thus the theorem is proved.

---

[1]) See Theorem 6.

[2]) Here and further below, $f': R \to R$ is defined by $f'(t)=f(t)$ for $t \in T$ and $f'(t)=0$ for $t \notin T$.

4. The notion of almost strong lifting

Let X be a locally compact space and $\mu \neq 0$ a positive Radon measure on X.

In order to cover the case where $\operatorname{Supp} \mu \neq X$ and in connection with some applications we shall also give the following:

Definition 5. – *Let ρ be a lifting of $M_R^\infty(X,\mu)$. We say that ρ is an almost strong lifting of $M_R^\infty(X,\mu)$ if there is a $\mu^\bullet$-negligible (that is, locally μ-negligible) set $A \subset X$ such that:*

$$\rho(f)| \complement A = f | \complement A$$

for all $f \in C_R^b(X)$.

It is easy to see that a lifting ρ of $M_R^\infty(X,\mu)$ is almost strong if and only if there is a $\mu^\bullet$-negligible set $A \subset X$ such that

$$\rho(f)| \complement A = f | \complement A$$

for every $f \in \mathscr{K}_R(X)$.

In the same way we could introduce (and discuss) the notion of *almost strong linear lifting*.

Remarks. – 1) Suppose that $\operatorname{Supp} \mu = X$. If ρ is a strong lifting of $M_R^\infty(X,\mu)$ then ρ is clearly an almost strong lifting. Conversely, let ρ be an almost strong lifting of $M_R^\infty(X,\mu)$. For each $t \in A$ (we use the notation of Definition 5) let χ_t be a character of $L_R^\infty(X,\mu)$ such that $\chi_t(\tilde{f}) = f(t)$ for all $f \in C_R^b(X)$ (see Proposition 1). For each $f \in M_R^\infty(X,\mu)$ define $\rho'(f)$ by

$$\rho'(f)(t) = \begin{cases} \rho(f)(t) & \text{if} \quad t \in \complement A, \\ \chi_t(\tilde{f}) & \text{if} \quad t \in A. \end{cases}$$

It is clear that ρ' is a strong lifting of $M_R^\infty(X,\mu)$ and that

$$\rho'(f)| \complement A = \rho(f)| \complement A$$

for all $f \in M_R^\infty(X,\mu)$.

It follows in particular[1]) that (X,μ) *has the strong lifting property if and only if there exists an almost strong lifting of* $M_R^\infty(X,\mu)$.

2) Notice that the Example of section 3 shows that in general, a lifting is *not* necessarily an almost strong lifting[2]). Therefore it is not necessarily true (for arbitrary X and $\mu \neq 0$) that a lifting of $M_R^\infty(X,\mu)$ can be "modified on a set of measure zero" so as to become strong.

---

[1]) Recall that $\operatorname{Supp} \mu = X$.

[2]) See statement 7.2) of Theorem 7.

Proposition 12. – *Let X be a locally compact space, $\mu \neq 0$ a positive Radon measure on X and suppose that $X_1 = \operatorname{Supp}\mu \neq X$. Suppose that ρ_1 is a strong lifting of $M_{\mathbf{R}}^{\infty}(X_1, \mu_{X_1})$. There is then an almost strong lifting ρ of $M_{\mathbf{R}}^{\infty}(X,\mu)$ such that:*

$$\rho(f)|X_1 = \rho_1(f|X_1) \quad \textit{for all} \quad f \in M_{\mathbf{R}}^{\infty}(X,\mu).$$

Proof: Let χ be a character of $L_{\mathbf{R}}^{\infty}(X,\mu)$. For $f \in M_{\mathbf{R}}^{\infty}(X,\mu)$ define $\rho(f)$ by

$$\rho(f)(t) = \begin{cases} \rho_1(f|X_1)(t) & \text{if} \quad t \in X_1, \\ \chi(\tilde{f}) & \text{if} \quad t \in \complement X_1. \end{cases}$$

It is clear that $\rho: f \to \rho(f)$ is a lifting of $M_{\mathbf{R}}^{\infty}(X,\mu)$ with the desired properties and thus Proposition 12 is proved.

Theorem 8. – *Let X be a locally compact metrizable space and $\mu \neq 0$ a positive Radon measure on X. Then any lifting of $M_{\mathbf{R}}^{\infty}(X,\mu)$ is an almost strong lifting.*

Proof: Let

$$A = \{t \in X | \text{ there exists } f \in C_{\mathbf{R}}^{b}(X) \text{ such that } \rho(f)(t) \neq f(t)\}.$$

For each compact $K \subset X$ let H_K be a countable part of $C_{\mathbf{R}}^{b}(X)$ dense in $C_{\mathbf{R}}^{b}(X)$ for the topology of uniform convergence on K. Let

$$A'_K = \{t \in K | \text{ there exists } f \in H_K \text{ such that } \rho(f)(t) \neq f(t)\}$$

and let

$$A_K = A'_K \cup (K - \rho(K)).$$

Then A_K *is $\mu^{\bullet}$-negligible and* $K - A_K \subset K \cap \rho(K)$.

Let now $t \in K - A_K$ and $f \in C_{\mathbf{R}}^{b}(X)$. Let (f_n) be a sequence of elements of H_K which converges to f uniformly on K. Then $(\varphi_{\rho(K)}\rho(f_n))$ converges uniformly (on X) to $\varphi_{\rho(K)}\rho(f)$. But (since $t \in K - A_K$)

$$\varphi_{\rho(K)}\rho(f_n)(t) = \rho(f_n)(t) = f_n(t) \qquad \text{for all } n,$$

and

$$\varphi_{\rho(K)}\rho(f)(t) = \rho(f)(t).$$

Hence

$$\rho(f)(t) = \lim_n f_n(t) = f(t).$$

Since $f \in C_{\mathbf{R}}^{b}(X)$ was arbitrary we deduce that $t \notin A$, that is, $A \cap K \subset A_K$. Since K was arbitrary we deduce that A is $\mu^{\bullet}$-negligible. Since

$$\rho(f)(t) = f(t)$$

for $f \in C_{\mathbf{R}}^{b}(X)$ and $t \in \complement A$, we deduce that ρ is almost strong.

From Remark 1) following Definition 5 we deduce that *if X is locally compact and metrizable and $\mu \neq 0$ a positive Radon measure on X with* $\operatorname{Supp}\mu = X$, *then (X,μ) has the strong lifting property.*

5. The notions of almost strong and strong lifting for topological spaces

Let T be a separated topological space and let $C_{\mathbf{R}}^b(T)$ be the algebra of all bounded continuous functions $f\colon T \to R$.

Let $\mathscr{E}$ be the tribe of the Borel parts of T and consider the measure space $(T,\mathscr{E},\alpha)$.

We use here the notations and terminology of I), section 9, chapter 1. In particular we denote by $\mathscr{R}$ the set of all $\mathscr{E}$-measurable simple functions f on T for which $\alpha(\{x \mid f(x) \neq 0\}) < +\infty$.

We shall suppose that:

(1) $\alpha(K) < +\infty$ if $K \subset T$ is compact;

(2) there exists a sequence (K_n) of *metrizable compact* parts of T such that

$$\alpha(T - K_\infty) = 0$$

where $K_\infty = \bigcup_n K_n$.

Let N be the upper integral corresponding to the measure space $(T,\mathscr{E},\alpha)$. From our assumptions we deduce that every compact set $K \subset T$ is N-integrable and $N(\varphi_K) = \alpha(K)$. Moreover (see Proposition 1, chapter 1) every set belonging to $\mathscr{E}$ is $(N,\mathscr{R})$-measurable.

From (2) it follows that $T - K_\infty$ is N-negligible and that $N = \bar{N}$. We conclude that $(T,N,\mathscr{R})$ is *strictly localizable.*

In the rest of this section we shall write M^∞ for $M^\infty(T,N,\mathscr{R})$.

Notice that in the definitions of strong and almost strong liftings (Definitions 1 and 5) the local compactness of the space considered and the fact that the considered measure was a Radon measure, were not used. We may therefore introduce, in the setting of this section, the following:

Definition 6. – *Let ρ be a lifting of M^∞. We say that ρ is an almost strong lifting of M^∞ if there is an N-negligible set $A \subset T$ such that*

$$\rho(f) \mid \complement A = f \mid \complement A$$

for all $f \in C_{\mathbf{R}}^b(T)$.

In the same way we could introduce (and discuss) the notion of *almost strong linear lifting.*

We shall now give the:

Theorem 9. – *Every lifting ρ of M^∞ is almost strong.*

The proof of this theorem is very similar to that of Theorem 8. For completeness, however, we shall give the details.

Proof: Let ρ be a lifting of M^∞. Let

$$A=\{t\in X \mid \text{ there exists } f\in C_R^b(T) \text{ such that } \rho(f)(t)\neq f(t)\}.$$

To prove the theorem it is enough to show that the set A is N-negligible. For this it is enough to prove that $A\cap K_n$ is N-negligible for all n. In fact, then $A\cap K_\infty$ is N-negligible and we have

$$A=(A\cap K_\infty)\cup((A\cap(T-K_\infty)).$$

Let n be fixed; to establish that $A\cap K_n$ is N-negligible we reason as follows: Let H_n be a countable part of $C_R^b(K_n)$ dense in $C_R^b(K_n)$. Let

$$A'_n=\{t\in K_n \mid \text{ there exists } f\in H_n \text{ such that } \rho(f')(t)\neq f'(t)\}.$$

Here f' is defined by $f'(t)=0$ if $t\notin K_n$ and $f'(t)=f(t)$ if $t\in K_n$. Notice that $f'\in M^\infty$ (clearly f' is the pointwise limit of a sequence of functions in $\mathscr{R}$).

Let $A_n=A'_n\cup(K_n-\rho(K_n))$; then A_n *is N-negligible and*

$$K_n-A_n\subset K_n\cap\rho(K_n).$$

Let now $t\in K_n-A_n$ and $f\in C_R^b(T)$. Let (f_p) be a sequence of elements of H_n which converges to $f|K_n$ uniformly on K_n. Then (f'_p) converges uniformly to $\varphi_{K_n}f$ on T, whence $(\rho(f'_p))$ converges uniformly (on T) to $\varphi_{\rho(K_n)}\rho(f)$. But (since $t\in K_n-A_n$),

$$\rho(f'_p)(t)=f'_p(t)=f_p(t) \quad \text{for all } p,$$

and

$$\varphi_{\rho(K_n)}\rho(f)(t)=\rho(f)(t).$$

Hence

$$\rho(f)(t)=\lim_p f_p(t)=f(t).$$

Since $f\in C_R^b(T)$ was arbitrary we deduce that $t\notin A$, that is, $A\cap K_n\subset A_n$. Since n was arbitrary, A is N-negligible.

Therefore Theorem 9 is proved.

Suppose now that α satisfies the following supplementary condition:

$$\alpha(U)\neq 0 \tag{3}$$

if U is open and non-empty.

Note that in this case two continuous functions coincide *everywhere* $\Leftrightarrow$ they coincide *N-almost everywhere.*

We now introduce the:

Definition 7. – *Let ρ be a lifting of M^∞. We say that ρ is a strong lifting of M^∞ if*

$$\rho(f)=f$$

for every $f\in C_R^b(T)$.

In the same way we introduce (and discuss) the notion of strong linear lifting. Using (a variant of) Proposition 1 of this chapter we deduce:

Theorem 10. – *There exists a strong lifting of* M^∞.

We close this section with the following remark: If T is a Polish space and if $(T,\mathscr{E},\alpha)$ is a measure space such that $\alpha(T)<\infty$, then conditions (1) and (2) are satisfied ([14], chap. 9, p. 121, [49], p. 40).

Appendix. Borel liftings

Let Z be a locally compact space, countable at infinity[1]) and let $\mu\neq 0$ be a positive Radon measure on Z. We shall say that a lifting ρ of $M_R^\infty(Z,\mu)$ is a *Borel lifting* if $\rho(f)$ is Borel measurable for every function $f\in M_R^\infty(Z,\mu)$.

The problem as to whether or not there exists a Borel lifting of $M_R^\infty(Z,\mu)$ is open, even in the case when $Z=[0,1]$ and $\mu=$the Lebesgue measure on $[0,1]$ (see [60] and [105] where the problem is partially solved). We now prove the:

Theorem A. – *Let Z be a locally compact space, countable at infinity, and $\mu\neq 0$ a positive Radon measure on Z. Suppose that there exists an almost strong lifting of $M_R^\infty(Z,\mu)$. Let (f_n) be a sequence of functions belonging to $M_R^\infty(Z,\mu)$ and let $\mathscr{A}$ be the closed[2]) subalgebra of $M_R^\infty(Z,\mu)$ spanned by $C_R^b(Z)$ and by the functions in the sequence (f_n). Then there exists an almost strong lifting ρ of $M_R^\infty(Z,\mu)$ such that $\rho(f)$ is Borel measurable for every $f\in\mathscr{A}$.*

Proof: Let δ be an almost strong lifting of $M_R^\infty(Z,\mu)$. For each n let $g_n\in M_R^\infty(Z,\mu)$ be a Borel measurable function and A_n a $\mu^\bullet$-negligible Borel set such that

$$\{z|\delta(f_n)(z)\neq g_n(z)\}\subset A_n.$$

Since δ is an almost strong lifting of $M_R^\infty(Z,\mu)$, there exists a $\mu^\bullet$-negligible Borel set B such that

$$\bigcup_{f\in C_R^b(Z)}\{z|\delta(f)(z)\neq f(z)\}\subset B.$$

Let

$$A=\Big(\bigcup_n A_n\Big)\cup B$$

---

[1]) That is, Z is a countable union of compact sets.

[2]) Closed for the "supremum norm".

and let χ be a character of $L^\infty_{\boldsymbol{R}}(Z,\mu)$. For every $f \in M^\infty_{\boldsymbol{R}}(Z,\mu)$ define $\rho(f)$ by the equations:

$$\rho(f)(z)=\begin{cases}\delta(f)(z) & \text{if } z \notin A,\\ \chi(\tilde{f}) & \text{if } z \in A.\end{cases}$$

Clearly $\rho: f \to \rho(f)$ is an almost strong lifting of $M^\infty_{\boldsymbol{R}}(Z,\mu)$. Moreover, if $f \in C^b_{\boldsymbol{R}}(Z)$ or $f=f_n$ for some n, then $\rho(f)$ is Borel measurable. Hence $\rho(f)$ is Borel measurable if $f \in \mathscr{A}$, and Theorem A is completely proved.

CHAPTER IX

Domination of measures and disintegration of measures

Throughout this chapter the setting is that of compact or locally compact spaces and positive Radon measures. We make constant use of the notation and terminology of chapter 8.

In section 1 we give a general theorem concerning convex cones of continuous functions and the domination of measures. In section 2 we give a proof of the disintegration theorem, in the particular case of a compact space and a continuous mapping, based on Strassen's theorem; we also show that, properly formulated ,"the disintegration of measures" is equivalent to "the existence of a strong lifting". Section 3 contains a discussion of the cones $\mathscr{F}(T,\mathscr{M}_+(S),\mu)$ and $\mathscr{F}^\infty(T,\mathscr{M}_+(S),\mu)$ of measure-valued mappings. In section 4 we introduce the notion of (measure-valued) mapping appropriate with respect to (μ,ρ) and we study the integrals of such mappings. The general form of the disintegration theorem is given in section 5; the proof is based on the Dunford-Pettis theorem and is independent of the particular form of the disintegration theorem given in section 2.

There are no separability assumptions throughout this chapter; the notion of strong lifting or almost strong lifting permits removing the separability assumptions in all the integral representation theorems.

We would like to point out that all the integral formulas in this chapter must be interpreted as "weak integrals".

Throughout this chapter, if X is a locally compact space (respectively a compact space) we write $\mathscr{K}(X)$ (respectively $C(X)$) instead of $\mathscr{K}_R(X)$ (respectively $C_R(X)$). For any locally compact space X, $\mathscr{M}_R(X)$ is the vector space of all real-valued Radon measures on X and $\mathscr{M}^b_R(X)$ is the vector space of all bounded real-valued Radon measures on X; again we shall simply write $\mathscr{M}(X)$ and $\mathscr{M}^b(X)$ instead of $\mathscr{M}_R(X)$ and $\mathscr{M}^b_R(X)$, respectively. We shall also use the notation $\mathscr{M}^1_+(X)=\{\mu\in\mathscr{M}^b_+(X)|\mu(1)=1\}$.

We recall that if X is a locally compact space and μ a positive Radon measure on X, then $(X,\mu^\bullet,\mathscr{K}(X))$ is strictly localizable. The equivalence relations "$\equiv$" considered here are of course taken "with respect to $\mu^\bullet$".

1. Convex cones of continuous functions and the domination of measures

Let S and T be two compact spaces and $\pi: S \to T$ a continuous mapping. Let $\mathscr{E} \subset C(T)$ be a convex cone[1]) such that:

a) $1 \in \mathscr{E}$;

b) $f \in \mathscr{E}$ *and* $g \in \mathscr{E} \Rightarrow \inf(f,g) \in \mathscr{E}$.

Let $a \in \pi(S)$ and let $p_a: C(S) \to R$ be defined by

$$p_a(f) = \inf\{g(a) \mid g \in \mathscr{E}, g \circ \pi \geqslant f\}.$$

Note that p_a is defined by the above relation only for $a \in \pi(S)$.

Remarks. – 1) It is easy to see that p_a is a *support function*[2]) on $C(S)$ and that

$$\langle g \circ \pi, p_a \rangle = g(a) \quad \text{for all} \quad g \in \mathscr{E}.$$

2) If $\gamma \in (C(S))' = \mathscr{M}(S)$ is dominated by p_a, then γ is *positive.*

In fact let $f \in C(S)$, $f \leqslant 0$. Then, using a) it is easily seen that $p_a(f) \leqslant 0$, whence $\gamma(f) \leqslant 0$.

3) If $\gamma \in (C(S))' = \mathscr{M}(S)$ is dominated by p_a, then

$$\langle g \circ \pi, \gamma \rangle = g(a) \quad \text{for all} \quad g \in \mathscr{E} \cap (-\mathscr{E}).$$

In fact if $g \in \mathscr{E}$, then by Remark 1),

$$\langle g \circ \pi, \gamma \rangle \leqslant \langle g \circ \pi, p_a \rangle = g(a).$$

4) If in particular we take $\mathscr{E} = C(T)$ and if $\gamma \in (C(S))' = \mathscr{M}(S)$ is *dominated by* p_a, then we obtain

$$\langle g \circ \pi, \gamma \rangle = g(a)$$

for all $g \in C(T)$, that is,

$$\pi(\gamma) = \varepsilon_a.$$

This shows that (see [15], chap. 5, p. 70)

$$\operatorname{Supp} \gamma \subset \pi^{-1}(\{a\}).$$

5) Let $K = \pi(S)$ and let $p_t = 0$ for $t \notin K$. Then $p: t \to p_t$ is a mapping of T into the set of support functions on $C(S)$. For each $f \in C(S)$ we have

$$\langle f, p \rangle = \inf\{\varphi_K g \mid g \in \mathscr{E}, g \circ \pi \geqslant f\}.$$

---

[1]) We recall that if L is a vector space, then $C \subset L$ is a convex cone if $\lambda C \subset C$ for every $\lambda > 0$ and $C + C \subset C$.

[2]) We use here the terminology and notation of section 6, chapter 7.

It follows in particular that $\langle f,p\rangle$ is *universally measurable*[1]) *and bounded.* Note also that the set

$$\{\varphi_K g \mid g\in\mathscr{E}, g\circ\pi\geqslant f\}$$

is directed (for the relation $\geqslant$). Hence if $\delta\in\mathscr{M}_+(T)$ we have

$$\int_T\langle f,p\rangle d\delta=\int_K(\langle f,p\rangle|K)d\delta_K=\inf\left\{\int_K(g|K)d\delta_K|g\in\mathscr{E},\ g\circ\pi\geqslant f\right\}$$
$$=\inf\left\{\int_T\varphi_K g\,d\delta|g\in\mathscr{E},\ g\circ\pi\geqslant f\right\}.$$

6) Let p be defined as in 5). Let $\lambda:t\to\lambda_t$ be a mapping of T into $\mathscr{M}(S)$. Then the following assertions are equivalent:

6.1) λ_t is dominated by p_t for every $t\in T$.

6.2) For each $t\in T$, $\lambda_t\in\mathscr{M}_+(S)$ and $\langle g\circ\pi,\lambda\rangle\leqslant\varphi_K g$ for every $g\in\mathscr{E}$.

It is enough to use Remarks 1) and 2) above.

Theorem 1 (Domination of measures). – *Let S and T be two compact spaces, $\pi:S\to T$ a continuous mapping and let $\mathscr{E}\subset C(T)$ be a convex cone satisfying* a) *and* b) *above. Let $\nu\in\mathscr{M}_+(S)$, $\mu\in\mathscr{M}_+(T)$, $\mu\neq 0$ such that $\operatorname{Supp}\mu\subset\pi(S)=K$. Suppose that ρ is an almost strong lifting of $M_R^\infty(T,\mu)$. Then the following assertions are equivalent:*

1.1) $\int_S g\circ\pi\,d\nu\leqslant\int_T g\,d\mu$ *for every* $g\in\mathscr{E}$.

1.2) *There is a mapping $\lambda:t\to\lambda_t$ of T into $\mathscr{M}_+(S)$ with the following properties:*

i) $\langle f,\lambda\rangle\in M_R^\infty(T,\mu)$ *for each* $f\in C(S)$;

ii) *There is a $\mu^\bullet$-negligible set $A\subset T$ such that* $\rho(\langle f,\lambda\rangle)(t)=\langle f,\lambda\rangle(t)$ *for every $f\in C(S)$ and $t\in\complement A$;*

iii) $\langle g\circ\pi,\lambda\rangle\leqslant\varphi_K g$ *for every* $g\in\mathscr{E}$;

iv) $\nu=\int_T\lambda_t\,d\mu(t)$[2]).

Proof: For each $t\in T$ define p_t on $C(S)$ by

$$p_t(f)=\inf\{\varphi_K(t)g(t)|g\in\mathscr{E},g\circ\pi\geqslant f\},\quad\text{for}\quad f\in C(S).$$

If we take $E=C(S)$ it is clear that $p:t\to p_t$ satisfies the hypotheses 11.1), 11.2) and 11.3) of Theorem 11, section 6, chapter 7. Conditions 11.1) and 11.2) follow from Remark 5) above. To verify 11.3) use the fact that ρ is an almost strong lifting. Since ρ is an almost strong lifting, there is a $\mu^\bullet$-negligible set $A\subset T$ such that $\rho(f)|\complement A=f|\complement A$ for every $f\in C(T)$. We may also suppose that $\rho(\varphi_K)(t)=\varphi_K(t)$ for $t\notin A$.

---

[1]) Clearly $\langle f,p\rangle|K$ is upper semicontinuous.

[2]) By this notation we mean that $\nu(f)=\int_T\langle f,\lambda_t\rangle d\mu(t)$ for each $f\in C(S)$.

Let q be the support function on $C(S)$ defined by

$$\langle f,q\rangle = \int_T \langle f,p_t\rangle \, d\mu(t), \quad \text{for} \quad f\in C(S).$$

We shall show that 1.1) is equivalent with:

α) The measure ν is dominated by q on $E=C(S)$.

1.1) ⇒ α). Let $f\in C(S)$. For every $g\in\mathscr{E}$ with $g\circ\pi \geqslant f$ we have

$$\int_S f\,d\nu \leqslant \int_S g\circ\pi \, d\nu \leqslant \int_T g\, d\mu.$$

Using Remark 5) above we deduce[1])

$$\int_S f\,d\nu \leqslant \int_T \langle f,p\rangle \, d\mu.$$

α) ⇒ 1.1). Let $g\in\mathscr{E}$. Since $\langle g\circ\pi,p\rangle = \varphi_K g$ (see Remark 1) above) we have

$$\int_S g\circ\pi\,d\nu \leqslant \int_T \langle g\circ\pi,p\rangle \, d\mu = \int_T g\,d\mu.$$

Thus the implication α) ⇒ 1.1) is also proved.

The equivalence of 1.1) and 1.2) now follows directly from Theorem 11, section 6, chapter 7 if we make use of Remark 6) above.

When $X=S=T$ is a compact convex metrizable set in a locally convex space, $\pi: X\to X$ is the identity mapping and $\mathscr{E}$ is the cone of continuous concave functions on X, Theorem 1 yields the well known theorem of Blackwell-Stein-Sherman-Cartier-Fell-Meyer (see [18] and [96], chapter 9).

2. Disintegration of measures. The case of a compact space and a continuous mapping

This important special case of the disintegration theorem can now be stated as follows:

Theorem 2. – *Let S and T be two compact spaces, $\pi: S\to T$ a continuous mapping and $K=\pi(S)$. Let $\nu\neq 0$ be a positive Radon measure on S and let $\mu=\pi(\nu)$. Suppose that ρ is an almost strong lifting of $M_{\mathbf{R}}^{\infty}(T,\mu)$.*

There is then a mapping $\lambda: t\to\lambda_t$ of T into $\mathscr{M}_+(S)$ with the following properties:

j) $\langle f,\lambda\rangle \in M_{\mathbf{R}}^{\infty}(T,\mu)$ for each $f\in C(S)$;

jj) There is a $\mu^\bullet$-negligible set $A\subset T$ such that $\rho(\langle f,\lambda\rangle)(t) = \langle f,\lambda\rangle(t)$ for each $f\in C(S)$ and $t\in \complement A$;

---

[1]) Use the fact that $\operatorname{Supp}\mu\subset K$, whence $\mu=\varphi_K\cdot\mu$.

jjj) $\langle g\circ\pi,\lambda\rangle=\varphi_K g$ *for every* $g\in C(T)$ *and hence* $\operatorname{Supp}\lambda_t\subset\pi^{-1}(t)$ *for every* $t\in T$;

jv) $\nu=\int_T \lambda_t\,d\mu(t)$.

Proof: Let $\mathscr{E}=C(T)$. Since $\mu=\pi(\nu)$, we have $\operatorname{Supp}\mu\subset\pi(S)$ and

$$\int_S g\circ\pi\,d\nu=\int_T g\,d\mu$$

for each $g\in\mathscr{E}$. Hence 1.1) of Theorem 1 is satisfied. By Theorem 1 there exists a mapping $\lambda:t\to\lambda_t$ of T into $\mathscr{M}_+(S)$ having the properties j), jj) and jv). This mapping λ satisfies also the relation

$$\langle g\circ\pi,\lambda\rangle\leqslant\varphi_K g$$

for every $g\in\mathscr{E}$. Since $\mathscr{E}$ is a vector space we deduce

$$\langle g\circ\pi,\lambda\rangle=\varphi_K g$$

for every $g\in\mathscr{E}$. Hence λ satisfies also jjj), and therefore the theorem is completely proved.

Remarks. – Let $\lambda:t\to\lambda_t$ be a mapping of T into $\mathscr{M}_+(S)$ such that: 1° $\langle f,\lambda\rangle\in M_R^\infty(T,\mu)$ for each $f\in C(S)$; 2° $\operatorname{Supp}\lambda_t\subset\pi^{-1}(t)$, $\mu^\bullet$-almost everywhere; 3° $\nu=\int_T\lambda_t\,d\mu(t)$. Then

$$\int_S f(g\circ\pi)\,d\nu=\int_T g\langle f,\lambda_t\rangle\,d\mu(t)$$

for all $f\in C(S)$ and $g\in C(T)$.

2) The mapping $\lambda:t\to\lambda_t$ of Theorem 2 is *unique* $\mu^\bullet$-almost everywhere, in the following sense: Let $\lambda':t\to\lambda'_t$ and $\lambda'':t\to\lambda''_t$ be two mappings of T into $\mathscr{M}_+(S)$ satisfying j), jj) (with $\mu^\bullet$-negligible sets $A'\subset T$ and $A''\subset T$, respectively), jjj) and jv) of Theorem 2. Then $\lambda'_t=\lambda''_t$, $\mu^\bullet$-almost everywhere.

In fact, by Remark 1) above, we have for all $f\in C(S)$ and $g\in C(T)$

$$\int_S f(g\circ\pi)\,d\nu=\int_T g\langle f,\lambda'\rangle\,d\mu$$

and

$$\int_S f(g\circ\pi)\,d\nu=\int_T g\langle f,\lambda''\rangle\,d\mu.$$

Thus

$$\int_T g\langle f,\lambda'\rangle\,d\mu=\int_T g\langle f,\lambda''\rangle\,d\mu$$

and hence (since $g \in C(T)$ was arbitrary), $\langle f, \lambda' \rangle \equiv \langle f, \lambda'' \rangle$. For $t \notin A' \cup A''$ we deduce

$$\langle f, \lambda'_t \rangle = \rho(\langle f, \lambda' \rangle)(t) = \rho(\langle f, \lambda'' \rangle)(t) = \langle f, \lambda''_t \rangle$$

for all $f \in C(S)$, that is, $\lambda'_t = \lambda''_t$ and the assertion is proved.

We next show that in a certain sense "the disintegration of measures" is equivalent to "the existence of a strong lifting".

Theorem 3. – *Let T be a compact space and μ a positive Radon measure on T with* $\operatorname{Supp} \mu = T$. *Then the following assertions are equivalent:*

3.1) *The couple (T, μ) has the strong lifting property.*

3.2) *For any $\{S, \nu, \pi\}$, where S is a compact space, ν a positive Radon measure on S and $\pi: S \to T$ a continuous mapping of S onto T such that $\pi(\nu) = \mu$, there is a mapping $\lambda: t \to \lambda_t$ of T into $\mathcal{M}_+^1(S)$ such that:*

i) $\nu = \int_T \lambda_t \, d\mu(t)$.

ii) $\operatorname{Supp} \lambda_t \subset \pi^{-1}(t)$ *for each* $t \in T$.

Proof: The implication 3.1) $\Rightarrow$ 3.2) follows from Theorem 2. We shall now prove the implication 3.2) $\Rightarrow$ 3.1). We shall use the fact that there exists a hyperstonean space (S, ν) and an isomorphism U of the Banach algebra $L_R^\infty(T, \mu)$ onto the Banach algebra $L_R^\infty(S, \nu)$ such that

$$\int_T f \, d\mu = \int_S U(\tilde{f}) \, d\nu$$

for every $f \in C(T)$.

Let ρ be the strong lifting of $M_R^\infty(S, \nu)$, which associates with every function $g \in M_R^\infty(S, \nu)$, the unique continuous function belonging to the class of g. For each $s \in S$ the mapping $f \to \rho(U(\tilde{f}))(s)$, of $C(T)$ into R, is a character of $C(T)$. Hence there exists a unique element $\pi(s) \in T$ such that

$$f(\pi(s)) = \rho(U(\tilde{f}))(s)$$

for all $f \in C(T)$. In this way we defined a mapping[1] $\pi: S \to T$. It is immediate that π is continuous and that $\mu = \pi(\nu)$.

By 3.2) there is then a mapping $\lambda: t \to \lambda_t$ of T into $\mathcal{M}_+^1(S)$ satisfying i) and ii). Define

$$\gamma(f) = \langle \rho(U(\tilde{f}), \lambda \rangle \quad \text{for} \quad f \in M_R^\infty(T, \mu).$$

Then $\gamma: f \to \gamma(f)$ is a strong linear lifting of $M_R^\infty(T, \mu)$. We show first that γ is a linear lifting. Clearly γ satisfies (II), (III), (IV), (V). To verify (I),

---

[1]) Since the mapping $f \to \rho(U(\tilde{f}))$ of $C(T)$ into $C(S)$ is injective, it follows that $\pi: S \to T$ is a surjection.

let $f\in M_R^\infty(T,\mu)$. Then $g=\rho(U(\tilde{f}))\in C(S)$. By i) and ii) we have for each $u\in C(T)$,

$$\int_S (u\circ\pi)g\,dv=\int_T u\langle g,\lambda\rangle d\mu=\int_T u\gamma(f)\,d\mu.$$

On the other hand, since $\tilde{u}\circ\pi=U(\tilde{u})$ and $\tilde{g}=U(\tilde{f})$ we also have

$$\int_S (u\circ\pi)g\,dv=\int_S U(\tilde{u})U(\tilde{f})dv=\int_S U(\tilde{u}\tilde{f})dv=\int_T u f\,d\mu.$$

We deduce

$$\int_T u\cdot\gamma(f)d\mu=\int_T u\cdot f\,d\mu \quad \text{for each} \quad u\in C(T),$$

whence $\gamma(f)\equiv f$. Thus (I) is proved, and γ is a linear lifting of $M_R^\infty(T,\mu)$ as asserted. To see that γ is strong, let $f\in C(T)$ and note that $\rho(U(\tilde{f}))=f\circ\pi$ and that $\langle f\circ\pi,\lambda\rangle=f$ (use ii)).

Since "the existence of a strong linear lifting of $M_R^\infty(T,\mu)$" is equivalent to "the existence of a strong lifting of $M_R^\infty(T,\mu)$", the implication 3.2)$\Rightarrow$3.1) is also proved.

Some of the methods used in this proof are similar to those of Chapter 10.

3. The cones $\mathscr{F}(T,\mathscr{M}_+(S),\mu)$ and $\mathscr{F}^\infty(T,\mathscr{M}_+(S),\mu)$

Let X be a locally compact space and $\mu\neq 0$ a positive Radon measure on X. We recall that we denoted by $\mathscr{C}(X,\mu)$ the set of all locally countable families $(K_j)_{j\in J}$ having the properties (a), (b), (c) of Definition 2, section 1, chapter 8.

Remarks. – 1) If $(K'_j)_{j\in J'}\in\mathscr{C}(X,\mu)$ and $(K''_j)_{j\in J''}\in\mathscr{C}(X,\mu)$ and if $H=\{(j',j'')\in J'\cap J''|\mu(K'_{j'}\cap K''_{j''})>0\}$ then

$$(K'_{j'}\cap K''_{j''})_{(j',j'')\in H}\in\mathscr{C}(X,\mu).$$

2) Let ρ be a lifting of $M_R^\infty(X,\mu)$. Then given $(K_j)_{j\in J}\in\mathscr{C}(X,\mu)$, there is $(L_\iota)_{\iota\in I}\in\mathscr{C}(X,\mu)$ such that each L_ι is contained in some $\rho(K_j)$.

Let now S *and* T *be two locally compact spaces.* For each mapping $\lambda: t\to\lambda_t$ of T into $\mathscr{M}_+(S)$ and for each $g\in\mathscr{K}(S)$ we denote by $\langle g,\lambda\rangle$ the mapping $t\to\langle g,\lambda_t\rangle$ of T into R.

Let $\mu\neq 0$ be *a positive Radon measure on* T.

Definition 1. – *We denote by* $\mathscr{F}(T,\mathscr{M}_+(S),\mu)$ *the convex cone of all mappings* $\lambda: t\to\lambda_t$ *of* T *into* $\mathscr{M}_+(S)$ *having the following property: There is* $(K_j)_{j\in J}\in\mathscr{C}(T,\mu)$ *such that*

$$\varphi_{K_j}\langle g,\lambda\rangle\in M_R^\infty(T,\mu)$$

for every $g\in\mathscr{K}(S)$ *and* $j\in J$.

Note that if $\lambda \in \mathscr{F}(T, \mathscr{M}_+(S), \mu)$ and $g \in \mathscr{K}(S)$, then $\langle g, \lambda \rangle$ is $\mu^\bullet$-measurable.

Let now ρ be *a lifting of* $M_R^\infty(T, \mu)$.

Definition 2. – *For $\lambda \in \mathscr{F}(T, \mathscr{M}_+(S), \mu)$ we shall write $\rho[\lambda] = \lambda$ whenever there is $(K_j)_{j \in J} \in \mathscr{C}(T, \mu)$ such that*

i) $\varphi_{K_j} \langle g, \lambda \rangle \in M_R^\infty(T, \mu)$ *for all* $g \in \mathscr{K}(S)$ *and* $j \in J$;

ii) $\rho(\varphi_{K_j} \langle g, \lambda \rangle) = \varphi_{\rho(K_j)} \langle g, \lambda \rangle$ *for all* $g \in \mathscr{K}(S)$ *and* $j \in J$.

For $\lambda' \in \mathscr{F}(T, \mathscr{M}_+(S), \mu)$ and $\lambda'' \in \mathscr{F}(T, \mathscr{M}_+(S), \mu)$ we shall write $\lambda' \equiv \lambda''$ (w)[1]) if $\langle g, \lambda' \rangle \equiv \langle g, \lambda'' \rangle$ for each $g \in \mathscr{K}(S)$. In this way we define an equivalence relation in $\mathscr{F}(T, \mathscr{M}_+(S), \mu)$.

The next result is similar to Proposition 3, section 6, chapter 6. For completeness we prove it in detail.

Proposition 1. – 1) *For every $\lambda \in \mathscr{F}(T, \mathscr{M}_+(S), \mu)$ there is*

$$\lambda' \in \mathscr{F}(T, \mathscr{M}_+(S), \mu)$$

such that $\rho[\lambda'] = \lambda'$ and $\lambda' \equiv \lambda$.

2) *If λ' and λ'' are in $\mathscr{F}(T, \mathscr{M}_+(S), \mu)$, $\rho[\lambda'] = \lambda'$, $\rho[\lambda''] = \lambda''$ and $\lambda' \equiv \lambda''$ then λ'_t and λ''_t coincide $\mu^\bullet$-almost everywhere.*

3) *If $\lambda \in \mathscr{F}(T, \mathscr{M}_+(S), \mu)$ and $\rho[\lambda] = \lambda$ then $t \to \lambda_t^\bullet(1)$ is $\mu^\bullet$-measurable.*

Proof: 1) Let $(K_j)_{j \in J} \in \mathscr{C}(T, \mu)$ be such that

$$\varphi_{K_j} \langle g, \lambda \rangle \in M_R^\infty(T, \mu)$$

for all $g \in \mathscr{K}(S)$ and $j \in J$. For each $g \in \mathscr{K}(S)$ define

$$\lambda'_t(g) = \begin{cases} \rho(\varphi_{K_j} \langle g, \lambda \rangle)(t) & \text{if } t \in \rho(K_j) \\ 0 & \text{if } t \notin \bigcup\limits_{j \in J} \rho(K_j). \end{cases}$$

Since $\rho(K_{j'}) \cap \rho(K_{j''}) = \emptyset$ if $j' \neq j''$ it follows that $\lambda'_t(g)$ is well defined for $g \in \mathscr{K}(S)$. The properties of ρ imply that λ'_t is a positive linear mapping of $\mathscr{K}(S)$ into R. Thus $\lambda' : t \to \lambda'_t$ is a mapping of T into $\mathscr{M}_+(S)$.

We shall now show that $\lambda' \in \mathscr{F}(T, \mathscr{M}_+(S), \mu)$, $\lambda' \equiv \lambda$ and $\rho[\lambda'] = \lambda'$.

Let $(L_i)_{i \in I} \in \mathscr{C}(T, \mu)$ be such that each L_i is contained in some $\rho(K_j)$ (see Remark 2) at the beginning of this section). Let $i \in I$ and $j \in J$ such that $L_i \subset \rho(K_j)$ and let $g \in \mathscr{K}(S)$; then

$$\varphi_{L_i} \langle g, \lambda' \rangle = \varphi_{L_i} \rho(\varphi_{K_j} \langle g, \lambda \rangle) \in M_R^\infty(T, \mu).$$

Since $g \in \mathscr{K}(S)$ and $i \in I$ were arbitrary we deduce $\lambda' \in \mathscr{F}(T, \mathscr{M}_+(S), \mu)$. Note now that for $i \in I, j \in J$ as above and $g \in \mathscr{K}(S)$ we have

$$\varphi_{L_i} \langle g, \lambda' \rangle = \varphi_{L_i} \rho(\varphi_{K_j} \langle g, \lambda \rangle) \equiv \varphi_{L_i} \varphi_{K_j} \langle g, \lambda \rangle \equiv \varphi_{L_i} \langle g, \lambda \rangle.$$

Since $i \in I$ was arbitrary, $\langle g, \lambda' \rangle \equiv \langle g, \lambda \rangle$; since g was also arbitrary it follows that $\lambda \equiv \lambda'$. Let again $i \in I$ and $j \in J$ as above and let $g \in \mathscr{K}(S)$.

---

[1]) When there is no ambiguity we simply write $\lambda' \equiv \lambda''$.

Then, using the fact that for $t\in\rho(K_j)$, $\langle g,\lambda'_t\rangle=\rho(\varphi_{K_j}\langle g,\lambda\rangle)(t)$, and the inclusion $L_i\subset\rho(K_j)$ (which obviously implies $\rho(L_i)\subset\rho(K_j)$) we deduce

$$\begin{aligned}\rho(\varphi_{L_i}\langle g,\lambda'\rangle)&=\rho(\varphi_{L_i}\varphi_{K_j}\langle g,\lambda'\rangle)=\rho(\varphi_{L_i}\varphi_{K_j}\langle g,\lambda\rangle)\\&=\varphi_{\rho(L_i)}\rho(\varphi_{K_j}\langle g,\lambda\rangle)=\varphi_{\rho(L_i)}\langle g,\lambda'\rangle\end{aligned}$$

and hence ($g\in\mathscr{K}(S)$ and $i\in I$ being arbitrary) $\lambda'=\rho[\lambda']$.

2) Let $(K'_j)_{j\in J'}\in\mathscr{C}(T,\mu)$ and $(K''_j)_{j\in J''}\in\mathscr{C}(T,\mu)$ be such that

$$\varphi_{K'_{j'}}\langle g,\lambda'\rangle\in M_R^\infty(T,\mu),\qquad \varphi_{K''_{j''}}\langle g,\lambda''\rangle\in M_R^\infty(T,\mu)$$

and

$$\rho(\varphi_{K'_{j'}}\langle g,\lambda'\rangle)=\varphi_{\rho(K'_{j'})}\langle g,\lambda'\rangle,$$

$$\rho(\varphi_{K''_{j''}}\langle g,\lambda''\rangle)=\varphi_{\rho(K''_{j''})}\langle g,\lambda''\rangle,$$

for all $g\in\mathscr{K}(S)$ and $j'\in J'$, $j''\in J''$. Let now

$$H=\{(j',j'')\in J'\times J''\,|\,\mu(K'_{j'}\cap K''_{j''})>0\};$$

then (see Remark 1) at the beginning of this section)

$$(K'_{j'}\cap K''_{j''})_{(j',j'')\in H}\in\mathscr{C}(T,\mu).$$

For $(j',j'')\in H$ and $g\in\mathscr{K}(S)$ we deduce

$$\begin{aligned}\varphi_{\rho(K'_{j'}\cap K''_{j''})}\langle g,\lambda'\rangle&=\varphi_{\rho(K''_{j''})}\rho(\varphi_{K'_{j'}}\langle g,\lambda'\rangle)=\rho(\varphi_{K''_{j''}\cap K'_{j'}}\langle g,\lambda'\rangle)\\&=\rho(\varphi_{K'_{j'}\cap K''_{j''}}\langle g,\lambda''\rangle)=\varphi_{\rho(K'_{j'})}\rho(\varphi_{K''_{j''}}\langle g,\lambda''\rangle)\\&=\varphi_{\rho(K'_{j'}\cap K''_{j''})}\langle g,\lambda''\rangle,\end{aligned}$$

thus $\lambda'_t=\lambda''_t$ for every $t\in\rho(K'_{j'}\cap K''_{j''})$. Since $(K'_{j'}\cap K''_{j''})_{(j',j'')\in H}\in\mathscr{C}(T,\mu)$ we deduce that λ'_t and λ''_t coincide $\mu^\bullet$-almost everywhere.

3) Let $(K_j)_{j\in J}\in\mathscr{C}(T,\mu)$ be such that

$$\varphi_{K_j}\langle g,\lambda\rangle\in M_R^\infty(T,\mu)$$

and

$$\rho(\varphi_{K_j}\langle g,\lambda\rangle)=\varphi_{\rho(K_j)}\langle g,\lambda\rangle$$

for all $g\in\mathscr{K}(S)$ and $j\in J$. Let $(L_i)_{i\in I}\in\mathscr{C}(T,\mu)$ be such that each L_i is contained in some $\rho(K_j)$ (see Remark 2) at the beginning of this section). Let $i\in I$ and $j\in J$ such that $L_i\subset\rho(K_j)$. For $t\in L_i$ we have

$$\begin{aligned}\lambda_t^\bullet(1)&=\sup\{\langle f,\lambda_t\rangle\,|\,f\in\mathscr{K}(S),0\leqslant f\leqslant 1\}\\&=\sup\{\rho(\varphi_{K_j}\langle f,\lambda\rangle)(t)\,|\,f\in\mathscr{K}(S),0\leqslant f\leqslant 1\}.\end{aligned}$$

But the function

$$\sup\{\rho(\varphi_{K_j}\langle f,\lambda\rangle)\,|\,f\in\mathscr{K}(S),0\leqslant f\leqslant 1\}$$

is $\mu^\bullet$-measurable (see Theorem 3, section 3, chapter 3). It follows that the restriction of $t\to\lambda_t^\bullet(1)$ to each L_i is $\mu^\bullet$-measurable. Hence the mapping $t\to\lambda_t^\bullet(1)$ is $\mu^\bullet$-measurable and the proposition is proved.

We shall also need later the following somewhat technical result:

Proposition 2. – *Let $\lambda \in \mathscr{F}(T, \mathscr{M}_+(S), \mu)$ be such that $\rho[\lambda] = \lambda$ and let $(K_j)_{j\in J} \in \mathscr{C}(T, \mu)$ satisfy* i) *and* ii) *of Definition* 2. *Let $L \subset S$ be a compact and let $\lambda_L: t \to \lambda_{L,t}$ be defined by $\lambda_{L,t} = (\lambda_t)_L$ (=restriction of λ_t to L). We then have:*

2.1) *For each $u \in C(L)$ and $j \in J$, $\varphi_{K_j}\langle u, \lambda_L\rangle \in M_R^\infty(T, \mu)$. Moreover if $u \geqslant 0$, then*

$$\rho(\varphi_{K_j}\langle u, \lambda_L\rangle) \leqslant \varphi_{\rho(K_j)}\langle u, \lambda_L\rangle$$

for all $j \in J$.

2.2) *$\lambda_L: t \to \lambda_{L,t}$ belongs to $\mathscr{F}(T, \mathscr{M}_+(L), \mu)$.*

Proof: 2.1) Let $u \in C_+(L)$. Let $u': S \to R$ be defined by $u'(s) = u(s)$ for $s \in L$ and $u'(s) = 0$ for $s \notin L$. Then u' is upper semicontinuous and hence $u' = \inf\{h | h \in \mathscr{K}(S), h \geqslant u'\}$ (pointwise infimum). We then have for each $t \in T$

$$\langle u, \lambda_{L,t}\rangle = \int_S u' \, d\lambda_t = \inf\{\langle h, \lambda_t\rangle | h \in \mathscr{K}(S), h \geqslant u'\}$$

whence

$$\varphi_B\langle u, \lambda_L\rangle = \inf\{\varphi_B\langle h, \lambda\rangle | h \in \mathscr{K}(S), h \geqslant u'\}$$

for every $\mu^\bullet$-measurable set $B \subset T$. Let $j \in J$. Since

$$\rho(\varphi_{K_j}\langle h, \lambda\rangle) = \varphi_{\rho(K_j)}\langle h, \lambda\rangle$$

for every $h \in \mathscr{K}(S)$ we deduce (use Theorem 3, section 3, chapter 3) that $\varphi_{\rho(K_j)}\langle u, \lambda_L\rangle$ is $\mu^\bullet$-measurable; hence $\varphi_{K_j}\langle u, \lambda_L\rangle$ is $\mu^\bullet$-measurable and (being obviously bounded),

$$\varphi_{K_j}\langle u, \lambda_L\rangle \in M_R^\infty(T, \mu).$$

We also have

$$\rho(\varphi_{K_j}\langle u, \lambda_L\rangle) = \rho(\varphi_{\rho(K_j)}\langle u, \lambda_L\rangle) \leqslant \varphi_{\rho(K_j)}\langle u, \lambda_L\rangle$$

for each $j \in J$.

Since 2.2) is a consequence of 2.1), the proposition is proved.

Definition 3. – *We denote by $\mathscr{F}^\infty(T, \mathscr{M}_+(S), \mu)$ the convex cone of all mappings $\lambda: t \to \lambda_t$ of T into $\mathscr{M}_+^b(S)$ such that*

j) *$\langle g, \lambda\rangle$ is $\mu^\bullet$-measurable for each $g \in \mathscr{K}(S)$;*

jj) *The mapping $t \to \|\lambda_t\|$ is bounded.*

With the notation of Definition 3, section 4, chapter 6 (see also section 5, chapter 6) it is clear that $\mathscr{F}^\infty(T, \mathscr{M}_+(S), \mu)$ is a subset of $M_{E'}^\infty[E]$, where E = the space of continuous real-valued functions on S vanishing at ∞ and E' = the space of bounded real-valued Radon measures on S.

Definition 4. – *For $\lambda \in \mathscr{F}^\infty(T, \mathscr{M}_+(S), \mu)$ we shall write $\rho(\lambda) = \lambda$ whenever*

$$\rho(\langle g, \lambda\rangle) = \langle g, \lambda\rangle$$

for all $g \in \mathscr{K}(S)$.

The notation $\rho(\lambda)=\lambda$ introduced in Definition 4 above agrees with that of section 4, chapter 6 (see Definition 4 and Proposition 1).

It is clear that $\mathscr{F}^{\infty}(T,\mathscr{M}_+(S),\mu)\subset\mathscr{F}(T,\mathscr{M}_+(S),\mu)$ and that if $\lambda\in\mathscr{F}^{\infty}(T,\mathscr{M}_+(S),\mu)$ and $\rho(\lambda)=\lambda$, then $\rho[\lambda]=\lambda$.

From Proposition 1, section 4, chapter 6 we easily obtain the following

Proposition 3. – 1) *For every* $\lambda\in\mathscr{F}^{\infty}(T,\mathscr{M}_+(S),\mu)$ *there is a mapping* $\lambda'\in\mathscr{F}^{\infty}(T,\mathscr{M}_+(S),\mu)$ *such that* $\rho(\lambda')=\lambda'$ *and* $\lambda'\equiv\lambda$.

2) *If* $\lambda\in\mathscr{F}^{\infty}(T,\mathscr{M}_+(S),\mu)$ *and* $\rho(\lambda)=\lambda$ *then* $t\to\|\lambda_t\|$ *is* $\mu^{\bullet}$*-measurable.*

We shall now make several remarks concerning Theorem 2 of section 2. Note first that the mapping λ in the statement of this theorem belongs to the cone $\mathscr{F}^{\infty}(T,\mathscr{M}_+(S),\mu)$.

Suppose now that ρ is a *strong lifting* of $M_R^{\infty}(T,\mu)$.

Let $\lambda'\in\mathscr{F}^{\infty}(T,\mathscr{M}_+(S),\mu)$ such that $\rho(\lambda')=\lambda'$ and $\lambda'\equiv\lambda$. Then

jj') $$\rho(\langle f,\lambda'\rangle)(t)=\langle f,\lambda'_t\rangle$$

for each $f\in C(S)$ and $t\in T$.

Since ρ is a strong lifting of $M_R^{\infty}(T,\mu)$, we must have $\operatorname{Supp}\mu=T$. But $\operatorname{Supp}\mu\subset\pi(S)=K$, whence $K=\pi(S)=T$. From jjj) we deduce for each $g\in C(T)$

$$\langle g\circ\pi,\lambda'\rangle=\rho(\langle g\circ\pi,\lambda'\rangle)=\rho(\langle g\circ\pi,\lambda\rangle)=\rho(g)=g,$$

that is

$$\langle g\circ\pi,\lambda'\rangle=g;$$

therefore $\operatorname{Supp}\lambda'_t\subset\pi^{-1}(t)$ *and* $\lambda'_t(1)=1$ *for all* $t\in T$.

4. Integration of measures

Throughout this section *S and T are locally compact spaces,* $\mu\neq0$ *a positive Radon measure on T and* ρ *a lifting of* $M_R^{\infty}(T,\mu)$.

Definition 5. – *A mapping* $\lambda\colon t\to\lambda_t$ *of* T *into* $\mathscr{M}_+(S)$ *is called appropriate with respect to* (μ,ρ) *if:*

i) $\lambda\in\mathscr{F}(T,\mathscr{M}_+(S),\mu)$ *and* $\rho[\lambda]=\lambda$;

ii) *For each* $g\in\mathscr{K}(S)$, $\langle g,\lambda\rangle$ *is* $\mu^{\bullet}$*-integrable.*

Remark. – Let $\lambda\colon t\to\lambda_t$ be a mapping of T into $\mathscr{M}_+(S)$ satisfying ii) of Definition 5. Then we may define $\nu=\int_T\lambda_t\,d\mu(t)$ by[1])

$$\langle g,\nu\rangle=\int_T\langle g,\lambda_t\rangle\,d\mu(t)$$

for $g\in\mathscr{K}(S)$; clearly ν is a positive Radon measure on S.

---

[1]) By abuse of notation, the integral of a $\mu^{\bullet}$-integrable function $f\colon T\to R$ will still be denoted by $\int_T f\,d\mu$ or $\int_T f(t)\,d\mu(t)$, instead of $\int_T f\,d\mu_{(\mu^{\bullet},\mathscr{K}(T))}$.

A mapping $\lambda: t \to \lambda_t$ of T into $\mathscr{M}_+(S)$ is called *strictly μ-adequate* if:

j) The mapping $\lambda: t \to \lambda_t$ is vaguely $\mu^\bullet$-measurable;

jj) For each $g \in \mathscr{K}(S)$, $\langle g, \lambda \rangle$ is $\mu^\bullet$-integrable.

Proposition 4. – *Suppose that ρ is a strong lifting of $M_R^\infty(T,\mu)$. Let $\lambda: t \to \lambda_t$ be a strictly μ-adequate mapping of T into $\mathscr{M}_+(S)$. Then λ is appropriate with respect to (μ, ρ).*

Proof: Since $\lambda: t \to \lambda_t$ is strictly μ-adequate, λ satisfies j) above and therefore there is $(K_j)_{j \in J} \in \mathscr{C}(T,\mu)$ such that $\langle g, \lambda \rangle | K_j$ is continuous for each $g \in \mathscr{K}(S)$ and each $j \in J$. This shows in particular that $\lambda \in \mathscr{F}(T, \mathscr{M}_+(S), \mu)$.

Let now $g \in \mathscr{K}(S)$ and $j \in J$. Let $h \in \mathscr{K}(T)$ be such that $\langle g, \lambda \rangle | K_j = h | K_j$. Since ρ is a strong lifting, $\rho(K_j) \subset K_j$ (see Theorem 1, section 1, chapter 8) and hence we have

$$\rho(\varphi_{K_j} \langle g, \lambda \rangle) = \rho(\varphi_{K_j} h) = \varphi_{\rho(K_j)} h = \varphi_{\rho(K_j)} \langle g, \lambda \rangle.$$

Thus $\rho[\lambda] = \lambda$ and λ satisfies i) of Definition 5. Since ii) is verified by any strictly μ-adequate mapping, we deduce that λ is appropriate with respect to (μ, ρ). Thus Proposition 4 is proved.

Let now α be a *positive Radon measure on T* and let $\lambda: t \to \lambda_t$ be a mapping of T into $\mathscr{M}_+(S)$ such that $\langle g, \lambda \rangle$ is $\alpha^\bullet$-integrable for every $g \in \mathscr{K}(S)$. Let

$$\beta = \int_T \lambda_t \, d\alpha(t).$$

We shall say that λ *verifies condition* (C_α) if:

(C_α) *For every lower semicontinuous f: $S \to \bar{R}_+$, the mapping $t \to \int_S^* f(s)\, d\lambda_t(s)$ is $\alpha^\bullet$-measurable and*

$$\int_S^* f(s)\, d\beta(s) = \int_T^\bullet d\alpha(t) \int_S^* f(s)\, d\lambda_t(s).$$

Proposition 5. – *If the mapping $\lambda: t \to \lambda_t$ of T into $\mathscr{M}_+(S)$ is appropriate with respect to (μ, ρ) then λ verifies condition (C_α) for every positive Radon measure α on T, $\alpha \leqslant \mu$.*

Proof: Let α be a positive Radon measure on T, $\alpha \leqslant \mu$. Then there exists $h \in M_R^\infty(T,\mu)$ such that $0 \leqslant h \leqslant 1$ and $\alpha = h \cdot \mu$. We may and shall suppose that

$$\rho(h) = h.$$

Let $(K_j)_{j \in J} \in \mathscr{C}(T,\mu)$ be such that

$$\varphi_{K_j} \langle g, \lambda \rangle \in M_R^\infty(T,\mu)$$

and

$$\rho(\varphi_{K_j} \langle g, \lambda \rangle) = \varphi_{\rho(K_j)} \langle g, \lambda \rangle$$

for all $g\in\mathscr{K}(S)$ and $j\in J$. Let $\mathscr{F}(J)=$the set of all finite parts of J. For each $H\in\mathscr{F}(J)$ let

$$K_H=\bigcup_{j\in H}K_j$$

and note that we have

$$\rho(\varphi_{K_H}\langle g,\lambda\rangle)=\varphi_{\rho(K_H)}\langle g,\lambda\rangle$$

for every $g\in\mathscr{K}(S)$.

Let now $f\colon S\to\bar{R}_+$ be lower semicontinuous and let

$$\mathscr{K}_f=\{g\in\mathscr{K}(S)|0\leqslant g\leqslant f\}.$$

Then we have

$$\sup_{g\in\mathscr{K}_f,H\in\mathscr{F}(J)}\rho(\varphi_{K_H}\langle g,\lambda\rangle)(t)=\int_S^* f(s)\,d\lambda_t(s)$$

for every $t\in\bigcup_{j\in J}\rho(K_j)$. Since the complement of $\bigcup_{j\in J}\rho(K_j)$ is $\mu^{\bullet}$-negligible, we deduce (use Theorem 3, section 3, chapter 3) that $t\to\int_S^* f(s)\,d\lambda_t(s)$ is $\mu^{\bullet}$-measurable, *whence* $\alpha^{\bullet}$-*measurable*, and that

$$\begin{aligned}\int_T^{\bullet} d\alpha(t)\int_S^* f(s)\,d\lambda_t(s)&=\int_T^{\bullet}\Big(\int_S^* f(s)\,d\lambda_t(s)\Big)h(t)\,d\mu(t)\\&=\sup_{g\in\mathscr{K}_f,H\in\mathscr{F}(J)}\int_T\varphi_{\rho(K_H)}\langle g,\lambda\rangle h\,d\mu\\&=\sup_{g\in\mathscr{K}_f}\Big(\sup_{H\in\mathscr{F}(J)}\int_T\varphi_{\rho(K_H)}\langle g,\lambda\rangle h\,d\mu\Big)\\&\sup_{g\in\mathscr{K}_f}\int_T\langle g,\lambda\rangle h\,d\mu=\sup_{g\in\mathscr{K}_f}\int_T\langle g,\lambda\rangle\,d\alpha\\&=\sup_{g\in\mathscr{K}_f}\int_S g\,d\beta=\int_S^* f(s)\,d\beta(s).\end{aligned}$$

Therefore λ verifies condition (C_α) and the proposition is proved.

On the basis of Proposition 5 above, one can prove the following useful theorem (see [15], chap. 5, pp. 23–25 and the remarks following the statement of the theorem):

Theorem 4. – *Let $\lambda\colon t\to\lambda_t$ be a mapping of T into $\mathscr{M}_+(S)$ appropriate with respect to (μ,ρ) and let $v=\int_T\lambda_t\,d\mu(t)$. Let*[1])

$$A\subset S\quad\textit{be a } v^*\textit{-negligible set}$$

and

$$f\colon S\to R\quad\text{a } v^*\textit{-integrable function.}$$

---

[1]) If δ is a positive Radon measure on a locally compact space X we denote by δ^* the upper integral $f\to\int_X^* f\,d\delta$. We note that if $f\colon X\to\bar{R}$ is δ^*-integrable, then f is $\delta^{\bullet}$-integrable and the corresponding integrals of f are equal.

Then there is a $\mu^\bullet$-negligible set $H \subset T$ such that:

i) *For each $t \notin H$, A is λ_t^*-negligible.*

ii) *For each $t \notin H$, f is λ_t^*-integrable.*

iii) *The mapping $t \to \int_S f\,d\lambda_t$ (defined $\mu^\bullet$-almost everywhere on T) is $\mu^\bullet$-integrable.*

iv) $\int_S f\,d\nu = \int_T d\mu(t) \int_S f\,d\lambda_t$.

Note that in this theorem we suppose A to be ν^*-negligible, not $\nu^\bullet$-negligible; also we suppose f to be ν^*-integrable, not $\nu^\bullet$-integrable.

Remarks. – 1) The condition (C_α) was introduced by the authors in [71]. In [15], chap. 5 (second edition), p. 18, N. Bourbaki calls *α-pre-adequate* a mapping λ which satisfies (C_α). Notice that Proposition 5 above shows that if λ is *appropriate with respect to* (μ, ρ), then λ is *μ-adequate* in the sense of [15], chap. 5 (second edition), p. 18 – 19.

2) It is noted in [15], chap. 5, p. 19 that a strictly μ-adequate mapping is μ-adequate.

5. Disintegration of measures. The general case

Let X and Y be two locally compact spaces, and m a positive Radon measure on X. We recall that a mapping $\pi: X \to Y$ is called *m-proper* ([15], chap. 5, p. 69) if π is $m^\bullet$-measurable and if for each $f \in \mathscr{K}(Y)$, $f \circ \pi$ is m -integrable. If $\pi: X \to Y$ is m-proper, then one can define the *image measure* $\pi(m)$ on Y. For the basic properties of the image measure see [15], chap. 5, p. 68 – 75.

We shall now give four propositions (Propositions 6, 7, 8 and 9 below) which will be used in the proof of the main theorem of this section.

Throughout Propositions 6, 7, 8 *and* 9 *we suppose that:*

(1) *S and T are locally compact spaces;*

(2) *ν is a positive Radon measure on S;*

(3) *$p: S \to T$ is a ν-proper mapping;*

(4) *$\mu \neq 0$ is a positive Radon measure on T such that $p(\nu)$ is absolutely continuous with respect to μ (that is, $p(\nu) = \psi \cdot \mu$ where $\psi \geqslant 0$ is locally $\mu^\bullet$-integrable*[1]));

(5) *ρ is a lifting of $M_R^\infty(T, \mu)$;*

(6) *$\lambda: t \to \lambda_t$ is a mapping of T into $\mathscr{M}_+(S)$ which belongs to the set $\mathscr{F}(T, \mathscr{M}_+(S), \mu)$ and is such that $\rho[\lambda] = \lambda$;*

(7) $\int_S g(f \circ p)\,d\nu = \int_T f \langle g, \lambda \rangle\,d\mu$ *for every $g \in \mathscr{K}(S)$ and every $f \in \mathscr{K}(T)$.*

---

[1]) We recall that a mapping $u: T \to R$ is locally $\mu^\bullet$-integrable if for each $f \in \mathscr{K}(T)$, uf is $\mu^\bullet$-integrable.

Note that in (7) it is implicitly assumed that, for each $g \in \mathscr{K}(S)$, the function $\langle g, \lambda \rangle$ is locally $\mu^\bullet$-integrable.

We may now state and prove:

Proposition 6. – *We have:*
6.1) *λ is appropriate with respect to (μ, p) and $\nu = \int_T \lambda_t \, d\mu(t)$;*

6.2) *For every f: $T \to R$, bounded $\mu^\bullet$-measurable and with compact support, and every g: $S \to R$ bounded $\nu^\bullet$-measurable and with compact support, the mapping $t \to \int_S g \, d\lambda_t$ (defined $\mu^\bullet$-almost everywhere on T) is $\mu^\bullet$-integrable and*

$$\int_S g(f \circ p) \, d\nu = \int_T f(t) \left(\int_S g \, d\lambda_t \right) d\mu(t).$$

Proof: 6.1) Let $g \in \mathscr{K}_+(S)$ and let $\mathscr{K}_1 = \{ f \in \mathscr{K}(T) | 0 \leqslant f \leqslant 1 \}$. We have

$$\begin{aligned} \int_T^\bullet \langle g, \lambda \rangle \, d\mu &= \sup_{f \in \mathscr{K}_1} \int_T f \langle g, \lambda \rangle \, d\mu = \sup_{f \in \mathscr{K}_1} \int_S g(f \circ p) \, d\nu \\ &= \sup_{f \in \mathscr{K}_1} \int_S f \circ p \, d(g \cdot \nu) = \sup_{f \in \mathscr{K}_1} \int_T f \, dp(g \cdot \nu) = \int_S g \, d\nu \end{aligned}$$

and hence 6.1) is proved.

6.2) follows easily from (7) and 6.1) by approximating f and g with functions from $\mathscr{K}(T)$ and $\mathscr{K}(S)$ and using Theorem 4.

Proposition 7. – *Suppose that ρ is a strong lifting of $M_R^\infty(T, \mu)$. Let $L \subset S$ be a compact such that $p|L$ is continuous and let λ_L: $t \to \lambda_{L,t}$ be defined by $\lambda_{L,t} = (\lambda_t)_L$ (= restriction of λ_t to L) for each $t \in T$. Then*

$$\operatorname{Supp} \lambda_{L,t} \subset p^{-1}(t) \cap L$$

$\mu^\bullet$-almost everywhere.

Proof: Let $u \in \mathscr{K}(L)$, $f \in \mathscr{K}(T)$; let u': $S \to R$ be defined by $u'(s) = u(s)$ for $s \in L$ and $u'(s) = 0$ for $s \notin L$. Let $p_1 = p|L$. We have (use 6.2) of Proposition 6):

$$\begin{aligned} \int_L u(f \circ p_1) \, d\nu_L &= \int_S u'(f \circ p) \, d\nu = \int_T f(t) \left(\int_S u' \, d\lambda_t \right) d\mu(t) \\ &= \int_T f(t) \left(\int_L u \, d\lambda_{L,t} \right) d\mu(t) = \int_T f \langle u, \lambda_L \rangle \, d\mu. \end{aligned} \tag{8}$$

Since $p_1(\nu_L) \leqslant p(\nu)$, we deduce that $p_1(\nu_L)$ is also absolutely continuous with respect to μ; whence $p_1(\nu_L) = \psi_1 \cdot \mu$ where $\psi_1 \geqslant 0$ is locally $\mu^\bullet$-integrable. Let now $(K_j)_{j \in J} \in \mathscr{C}(T, \mu)$ be such that

$$\varphi_{K_j} \langle g, \lambda \rangle \in M_R^\infty(T, \mu)$$

and

$$\rho(\varphi_{K_j} \langle g, \lambda \rangle) = \varphi_{\rho(K_j)} \langle g, \lambda \rangle$$

for all $g \in \mathscr{K}(S)$ and $j \in J$. We may also assume without loss of generality that

(9) $\varphi_{K_j} \psi_1 \in M_R^\infty(T,\mu)$ and $\rho(\varphi_{K_j} \psi_1) = \varphi_{\rho(K_j)} \psi_1$ for all $j \in J$.

Let now $h \in \mathscr{K}(T)$. For each $f \in \mathscr{K}(T)$ we have

$$\int_L (h \circ p_1)(f \circ p_1)\, d\nu_L = \int_T hf\, dp_1(\nu_L) = \int_T hf\psi_1\, d\mu$$

and by (8)

$$\int_L (h \circ p_1)(f \circ p_1)\, d\nu_L = \int_T f \langle h \circ p_1, \lambda_L \rangle\, d\mu.$$

Since $f \in \mathscr{K}(T)$ was arbitrary we deduce

(10) $$\langle h \circ p_1, \lambda_L \rangle \equiv h\psi_1$$

for every $h \in \mathscr{K}(T)$. From (9), (10) and Proposition 2 we deduce for each $h \in \mathscr{K}_+(T)$

$$\varphi_{\rho(K_j)} h\psi_1 = \rho(\varphi_{K_j} h\psi_1) = \rho(\varphi_{K_j} \langle h \circ p_1, \lambda_L \rangle) \leqslant \varphi_{\rho(K_j)} \langle h \circ p_1, \lambda_L \rangle$$

for all $j \in J$, that is

$$h(t)\psi_1(t) \leqslant \langle h \circ p_1, \lambda_{L,t} \rangle = \langle h, p_1(\lambda_{L,t}) \rangle$$

for all $t \in \bigcup_{j \in J} \rho(K_j)$. Let $A = \complement\left(\bigcup_{j \in J} \rho(K_j)\right)$; then A is $\mu^\bullet$-negligible and we may rewrite the above inequalities under the form

(11) $$\psi_1(t)\varepsilon_t \leqslant p_1(\lambda_{L,t}) \quad \text{for each} \quad t \notin A.$$

Let $K = p_1(L)$. Then $K \subset T$ is a compact containing the support of $p_1(\lambda_{L,t})$ for each $t \in T$ and the support of $p_1(\nu_L) = \psi_1 \cdot \mu$; in particular we obtain that ψ_1 vanishes $\mu^\bullet$-almost everywhere outside K. From this observation and from (10) we deduce that $p_1(\lambda_{L,t})^\bullet(1) = \psi_1(t)$ for all $t \notin B$, where B is $\mu^\bullet$-negligible. Combining this last relation with (11) we obtain, for each $t \notin A \cup B$, $p_1(\lambda_{L,t}) = \psi_1(t)\varepsilon_t$, whence $\operatorname{Supp} \lambda_{L,t} \subset p_1^{-1}(t) = p^{-1}(t) \cap L$. Since $A \cup B$ is $\mu^\bullet$-negligible, the proposition is proved.

Proposition 8. – *The measure λ_t is bounded and $\|\lambda_t\| = \psi(t)$, $\mu^\bullet$-almost everywhere.*

Proof: Let $(K_j)_{j \in J} \in \mathscr{C}(T,\mu)$ be such that

$$\varphi_{K_j} \langle g, \lambda \rangle \in M_R^\infty(T,\mu)$$

and

$$\rho(\varphi_{K_j} \langle g, \lambda \rangle = \varphi_{\rho(K_j)} \langle g, \lambda \rangle$$

for all $g \in \mathscr{K}(S)$ and $j \in J$. Let $\mathscr{F}(J) =$ the set of all finite parts of J. For each $H \in \mathscr{F}(J)$ let

$$K_H = \bigcup_{j \in H} K_j.$$

Note that for every $H \in \mathscr{F}(J)$ and $g \in \mathscr{K}(S)$ we have

$$\rho(\varphi_{K_H}\langle g,\lambda\rangle) = \varphi_{\rho(K_H)}\langle g,\lambda\rangle.$$

Let now $f \in \mathscr{K}_+(T)$ and let $\mathscr{K}_1 = \{g \in \mathscr{K}(S) | 0 \leqslant g \leqslant 1\}$. Then $(\varphi_{\rho(K_H)}\langle g,\lambda\rangle)_{g \in \mathscr{K}_1, H \in \mathscr{F}(J)}$ is directed for the relation $\leqslant$ and its pointwise supremum is $\lambda_t^\bullet(1)$ for each $t \in \bigcup_{j \in J} \rho(K_j)$. By Theorem 3, section 3, chapter 3 (use also Proposition 6) we have:

$$\begin{aligned}\int_T f(t)\lambda_t^\bullet(1)\,d\mu(t) &= \sup_{g \in \mathscr{K}_1, H \in \mathscr{F}(J)} \int_T f\varphi_{\rho(K_H)}\langle g,\lambda\rangle\,d\mu \\ &= \sup_{g \in \mathscr{K}_1, H \in \mathscr{F}(J)} \int_S g((f\varphi_{K_H})\circ p)\,dv = \sup_{H \in \mathscr{F}(J)} \int_S (f\varphi_{K_H})\circ p\,dv \\ &= \sup_{H \in \mathscr{F}(J)} \int_T f\varphi_{K_H}\,dp(v) = \int_T f\,dp(v) = \int_T f\psi\,d\mu.\end{aligned}$$

Since $f \in \mathscr{K}_+(T)$ was arbitrary, we deduce that $\|\lambda_t\| = \psi(t) < +\infty$, $\mu^\bullet$-almost everywhere.

Thus Proposition 8 is proved.

Proposition 9. – *Suppose that v is bounded and that ρ is strong. Let $(L_n)_{n \in N} \in \mathscr{C}(S,v)$ be such that $p|L_n$ is continuous for each n. Let $A = \complement\left(\bigcup_n L_n\right)$. Then:*

9.1) *There is a $\mu^\bullet$-negligible set $H \subset T$ such that A is λ_t^*-negligible for each $t \notin H$.*

9.2) *λ_t is concentrated on $p^{-1}(t)$, $\mu^\bullet$-almost everywhere.*

Proof: The assertion 9.1) follows from Theorem 4, at the end of section 4, since A in this case is v^*-negligible.

9.2) By Propositions 7 and 8 there is a $\mu^\bullet$-negligible set $B \subset T$ such that

$$\lambda_t \text{ is bounded and } \operatorname{Supp} \lambda_{L_n,t} \subset p^{-1}(t) \cap L_n$$

for all n if $t \notin B$. Fix now $t \notin H \cup B$. Note that $p^{-1}(t) \cap L_n$ is compact for all n, that $p^{-1}(t) \cap A$ is $\lambda_t^\bullet$-negligible and hence that

$$p^{-1}(t) = \left(\bigcup_n p^{-1}(t) \cap L_n\right) \cup (p^{-1}(t) \cap A)$$

is $\lambda_t^\bullet$-measurable. Using the properties of the induced measure ([15], chapter 5, p. 82) we deduce

$$\begin{aligned}\lambda_t(p^{-1}(t)) &= \sum_n \lambda_t(p^{-1}(t) \cap L_n) + \lambda_t(p^{-1}(t) \cap A) \\ &= \sum_n \lambda_{L_n,t}(p^{-1}(t) \cap L_n) = \sum_n \lambda_{L_n,t}(L_n) = \sum_n \lambda_t(L_n) = \lambda_t\left(\bigcup_n L_n\right) = \lambda_t(S);\end{aligned}$$

since λ_t is bounded, this shows that λ_t is concentrated on $p^{-1}(t)$. Hence Proposition 9 is proved.

We are now in a position to state and prove the general form of the disintegration theorem:

Theorem 5 (Disintegration of measures). – *Let S and T be two locally compact spaces, ν a positive Radon measure on S, $p: S \to T$ a ν-proper mapping. Let $\mu \neq 0$ be a positive Radon measure on T such that $p(\nu)$ is absolutely continuous with respect to μ, that is $p(\nu) = \psi \cdot \mu$ for some $\psi \geqslant 0$ locally $\mu^\bullet$-integrable. Let ρ be a lifting of $M_R^\infty(T,\mu)$. Then:*

5.1) *There is a mapping $\lambda: t \to \lambda_t$ of T into $\mathcal{M}_+(S)$, appropriate with respect to (μ,ρ), such that:*

i) $\|\lambda_t\| = \psi(t)$, *$\mu^\bullet$-almost everywhere;*

ii) $\int_S g(f \circ p)\, d\nu = \int_T f \langle g, \lambda \rangle\, d\mu$ *for every $f \in \mathcal{K}(T)$ and $g \in \mathcal{K}(S)$.*

iii) *If in addition the lifting ρ is strong, then λ_t is concentrated on $p^{-1}(t)$, $\mu^\bullet$-almost everywhere.*

5.2) *Let $\lambda': t \to \lambda'_t$ and $\lambda'': t \to \lambda''_t$ be two mappings of T into $\mathcal{M}_+(S)$, appropriate with respect to (μ,ρ)[1]) and such that:*

j) *λ'_t and λ''_t are concentrated on $p^{-1}(t)$, $\mu^\bullet$-almost everywhere;*

jj) $\nu = \int_T \lambda'_t\, d\mu(t) = \int_T \lambda''_t\, d\mu(t)$.

Then $\lambda'_t = \lambda''_t$, $\mu^\bullet$-almost everywhere.

Proof: 5.1) Consider the mapping $f \to (f \circ p) \cdot \nu$ of $\mathcal{K}(T)$ into the Banach space $\mathcal{M}^b(S)$ endowed with the usual norm; recall that $\mathcal{M}^b(S)$ is the dual of the Banach space of all continuous real-valued functions on S vanishing at infinity. We have

$$\|(f \circ p) \cdot \nu\| = \int_S |f \circ p|\, d\nu = \int_T |f|\, dp(\nu).$$

We deduce that $f \to (f \circ p) \cdot \nu$ is a positive continuous linear mapping of $\mathcal{K}(T) \subset \mathcal{L}^1(T, p(\nu))$ into $\mathcal{M}^b(S)$. By the Dunford-Pettis theorem (Corollary 1 of Theorem 1, section 1, chapter 7) there is then a mapping $\delta: t \to \delta_t$ of T into $\mathcal{M}_+(S)$ belonging to $\mathcal{F}^\infty(T, \mathcal{M}_+(S), p(\nu))$ verifying

$$\int_S g(f \circ p)\, d\nu = \int_T f \langle g, \delta \rangle\, dp(\nu)$$

for all $g \in \mathcal{K}(S)$ and all $f \in \mathcal{K}(T)$. Define $\gamma_t = \psi(t)\delta_t$ for all $t \in T$. It is clear that $\gamma: t \to \gamma_t$ belongs to $\mathcal{F}(T, \mathcal{M}_+(S), \mu)$ and that[2])

$$\int_S g(f \circ p)\, d\nu = \int_T f \langle g, \gamma \rangle\, d\mu$$

for all $g \in \mathcal{K}(S)$ and all $f \in \mathcal{K}(T)$. By Proposition 1, there is $\lambda \in \mathcal{F}(T, \mathcal{M}_+(S), \mu)$ such that $\lambda \equiv \gamma$ and $\rho[\lambda] = \lambda$. Clearly λ satisfies ii).

---

[1]) Here the lifting ρ is not necessarily supposed to be strong.

[2]) Note that $\langle g, \gamma \rangle$ is locally $\mu^\bullet$-integrable for every $g \in \mathcal{K}(S)$.

By Proposition 6, λ is appropriate with respect to (μ,ρ) and by Proposition 8, λ satisfies also i). We shall now prove iii). Let $(K_j)_{j\in J}\in\mathscr{C}(T,\mu)$ be such that

$$\varphi_{K_j}\langle g,\lambda\rangle\in M_R^\infty(T,\mu) \quad\text{and}\quad \rho(\varphi_{K_j}\langle g,\lambda\rangle)=\varphi_{\rho(K_j)}\langle g,\lambda\rangle$$

for all $g\in\mathscr{K}(S)$ and $j\in J$. For each $j\in J$ and $t\in T$ define

$$\lambda_t^j=\varphi_{\rho(K_j)}(t)\lambda_t.$$

To establish iii) it is enough to show that, for each $j\in J$, λ_t^j is concentrated on $p^{-1}(t)$, $\mu^\bullet$-almost everywhere. In fact, suppose that this is true. For each $j\in J$, let H_j be the set of all $t\in T$ such that λ_t^j is not concentrated on $p^{-1}(t)$. Since $\lambda_t^j=0$ for $t\notin\rho(K_j)$, it follows that $H_j\subset\rho(K_j)$. Therefore

$$H_\infty=\bigcup_{j\in J}H_j$$

is $\mu^\bullet$-negligible. If $t\in\bigcup_{j\in J}\rho(K_j)$ and $t\notin H_\infty$, then λ_t $(=\lambda_t^j$ for some $j\in J)$ is concentrated on $p^{-1}(t)$.

Hence it remains to show that for each $j\in J$, λ_t^j is concentrated on $p^{-1}(t)$, $\mu^\bullet$-almost everywhere.

Fix $j\in J$ and consider $\lambda^j: t\to\lambda_t^j$. Since

$$\langle g,\lambda^j\rangle=\varphi_{\rho(K_j)}\langle g,\lambda\rangle=\rho(\varphi_{K_j}\langle g,\lambda\rangle)=\rho(\varphi_{\rho(K_j)}\langle g,\lambda\rangle)=\rho(\langle g,\lambda^j\rangle),$$

for each $g\in\mathscr{K}(S)$, it is clear that $\lambda^j\in\mathscr{F}(T,\mathscr{M}_+(S),\mu)$ and that $\rho[\lambda^j]=\lambda^j$. By 6.2) of Proposition 6 we have for all $g\in\mathscr{K}(S)$ and $f\in\mathscr{K}(T)$:

$$\int_S g((f\varphi_{K_j})\circ p)\,d\nu=\int_T f\varphi_{\rho(K_j)}\langle g,\lambda\rangle\,d\mu=\int_T f\langle g,\lambda^j\rangle\,d\mu.$$

Denote by ν_j the measure $(\varphi_{K_j}\circ p)\cdot\nu$ and note that $\nu_j\leqslant\nu$. With this notation we may rewrite the above formula as follows:

$$\int_S g(f\circ p)\,d\nu_j=\int_T f\langle g,\lambda^j\rangle\,d\mu$$

for all $g\in\mathscr{K}(S)$ and $f\in\mathscr{K}(T)$. Note that S, T, ν_j, p, μ, ρ, λ^j satisfy again (1)–(7). Since ν_j is bounded and ρ is strong, 9.2) of Proposition 9 can be used, and we conclude that λ_t^j is concentrated on $p^{-1}(t)$, $\mu^\bullet$-almost everywhere.

5.2) Let $A\subset T$ be the set of all $t\in T$ such that λ_t' is not concentrated on $p^{-1}(t)$; then A is $\mu^\bullet$-negligible. Let now $f\in\mathscr{K}(T)$ and $g\in\mathscr{K}(S)$; there is then a $\mu^\bullet$-negligible set $A_{f,g}\subset T$ such that $g(f\circ p)$ is $(\lambda_t')^\bullet$-integrable for every $t\notin A_{f,g}$ (note that $g(f\circ p)$ is ν^*-integrable since it has compact support and use ii) of Theorem 4). Then for $t\notin A\cup A_{f,g}$ we have:

$$\int_S g(f\circ p)\,d\lambda_t'=\int_S\varphi_{p^{-1}(t)}(s)g(s)f(p(s))\,d\lambda_t'(s)=f(t)\int_S g(s)\,d\lambda_t'(s)=f(t)\langle g,\lambda_t'\rangle.$$

We deduce by Theorem 4:

$$\int_S g(f\circ p)\,dv=\int_T d\mu(t)\int_S g(f\circ p)\,d\lambda_t'=\int_T f(t)\langle g,\lambda_t'\rangle\,d\mu(t).$$

In the same way we prove that

$$\int_S g(f\circ p)\,dv=\int_T f(t)\langle g,\lambda_t''\rangle\,d\mu(t)$$

for all $f\in\mathscr{K}(T)$ and $g\in\mathscr{K}(S)$.

The above formulas show that $\lambda'\equiv\lambda''$. Since $\rho[\lambda']=\lambda'$ and $\rho[\lambda'']=\lambda''$, we deduce from 2) of Proposition 1 that $\lambda_t'=\lambda_t''$, $\mu^\bullet$-almost everywhere. This concludes the proof of Theorem 5.

The mapping $\lambda: t\to\lambda_t$ in 5.1) of Theorem 5 is unique $\mu^\bullet$-almost everywhere in the following sense:

Proposition 10. – *Let $\lambda': t\to\lambda_t'$ and $\lambda'': t\to\lambda_t''$ be two mappings of T into $\mathscr{M}_+(S)$ appropriate with respect to (μ,ρ) and both satisfying* ii). *Then $\lambda_t'=\lambda_t''$, $\mu^\bullet$-almost everywhere.*

Proof: Let $g\in\mathscr{K}(S)$. By ii) we have

$$\int_T f\langle g,\lambda'\rangle\,d\mu=\int_T f\langle g,\lambda''\rangle\,d\mu$$

for all $f\in\mathscr{K}(T)$, whence $\langle g,\lambda'\rangle\equiv\langle g,\lambda''\rangle$. This shows that $\lambda'\equiv\lambda''$. Since moreover $\rho[\lambda']=\lambda'$ and $\rho[\lambda'']=\lambda''$, we deduce from 2) of Proposition 1 that $\lambda_t'=\lambda_t''$, $\mu^\bullet$-almost everywhere.

In the remarks that follow we write:

$$T_1=\operatorname{Supp}\mu \quad \textit{and} \quad \mu_1=\mu_{T_1} \quad (=\textit{restriction of } \mu \textit{ to } T_1).$$

Below we use the notation of Theorem 5. In order to cover the disintegration of measures in the case when the support of μ is not equal to T, we shall show that iii) *of* 5.1) *remains valid if we replace the assumption that ρ is strong by the assumption that ρ is almost strong.* For this purpose we need a series of remarks.

Remarks. – 1) Suppose that for each point $a\in T_1$, $\{a\}$ is not $\mu^\bullet$-negligible. Let $\lambda: t\to\lambda_t$ be a mapping of T into $\mathscr{M}_+(S)$ appropriate with respect to (μ,ρ) and satisfying ii) of 5.1). Then λ_t is concentrated on $p^{-1}(t)$ for every $t\in T_1$.

In fact, let $a\in T_1$. For each $g\in\mathscr{K}(S)$ we have (by 6.2) of Proposition 6)

$$\int_S g(\varphi_{\{a\}}\circ p)\,dv=\int_T \varphi_{\{a\}}(t)\Big(\int_S g\,d\lambda_t\Big)\,d\mu(t)=\Big(\int_S g\,d\lambda_a\Big)\,\mu(\{a\}),$$

$$\begin{aligned}\int_S g(\varphi_{\{a\}}\circ p)\,dv&=\int_T \varphi_{\{a\}}(t)\Big(\int_S(\varphi_{\{a\}}\circ p)g\,d\lambda_t\Big)\,d\mu(t)\\&=\Big(\int_S(\varphi_{\{a\}}\circ p)g\,d\lambda_a\Big)\,\mu(\{a\})\end{aligned}$$

whence

$$\int_S 1\cdot g\,d\lambda_a=\int_S(\varphi_{\{a\}}\circ p)g\,d\lambda_a=\int_S \varphi_{p^{-1}(a)}g\,d\lambda_a.$$

Since $g \in \mathcal{K}(S)$ was arbitrary, we deduce that $\varphi_{p^{-1}(a)}$ is equal to 1, $\lambda_a^\bullet$-almost everywhere and the assertion is proved.

2) Let ρ' and ρ'' be two liftings of $M_R^\infty(T,\mu)$ and suppose that there exists a $\mu^\bullet$-negligible set $A \subset T$ such that

$$\rho'(f)(t) = \rho''(f)(t)$$

for all $f \in M_R^\infty(T,\mu)$ and $t \notin A$. Let λ' and λ'' in $\mathcal{F}(T, \mathcal{M}_+(S), \mu)$ be such that $\lambda' \equiv \lambda''$,

$$\rho'[\lambda'] = \lambda' \quad \text{and} \quad \rho''[\lambda''] = \lambda''.$$

Then (as in the proof of 2) of Proposition 1) one can show that λ_t' and λ_t'' coincide $\mu^\bullet$-almost everywhere.

3) Let ρ' and ρ'' be two liftings of $M_R^\infty(T,\mu)$ and suppose that there exists a $\mu^\bullet$-negligible set $A \subset T$ such that

$$\rho'(f)(t) = \rho''(f)(t)$$

for all $f \in M_R^\infty(T,\mu)$ and $t \notin A$. Let $\lambda': t \to \lambda_t'$ and $\lambda'': t \to \lambda_t''$ be two mappings of T into $\mathcal{M}_+(S)$, appropriate with respect to (μ, ρ') and (μ, ρ'') respectively, and both satisfying ii) of 5.1). Then $\lambda_t' = \lambda_t''$, $\mu^\bullet$-almost everywhere.

It is enough to note that $\lambda' \equiv \lambda''$, on account of ii), and to use Remark 2).

4) Let ρ' be an almost strong lifting of $M_R^\infty(T,\mu)$. Then (see Proposition 12, section 4, chapter 8), there is a strong lifting ρ_1 of $M_R^\infty(T_1, \mu_1)$, an almost strong lifting ρ of $M_R^\infty(T,\mu)$ such that $\rho(f)|T_1 = \rho_1(f|T_1)$ for all $f \in M_R^\infty(T,\mu)$, and a $\mu^\bullet$-negligible set $A \subset T$ such that

$$\rho(f)(t) = \rho'(f)(t)$$

for all $f \in M_R^\infty(T,\mu)$ and $t \notin A$.

5) Let ρ_1 be a lifting of $M_R^\infty(T_1, \mu_1)$ and ρ a lifting of $M_R^\infty(T,\mu)$ such that

$$\rho(f)|T_1 = \rho_1(f|T_1)$$

for all $f \in M_R^\infty(T,\mu)$. Let $\lambda: t \to \lambda_t$ belong to $\mathcal{F}(T, \mathcal{M}_+(S), \mu)$ be such that $\rho[\lambda] = \lambda$. Let $\lambda_1 = \lambda|T_1$. Then the mapping $\lambda_1: t \to \lambda_t$ of T_1 into $\mathcal{M}_+(S)$ belongs to $\mathcal{F}(T_1, \mathcal{M}_+(S), \mu_1)$ and $\rho_1[\lambda_1] = \lambda_1$.

6) Consider the ν-proper mapping $p: S \to T$. Then $N = T - T_1$ is $(p(\nu))^\bullet$-negligible (since it is $\mu^\bullet$-negligible); hence $B = p^{-1}(N)$ is $\nu^\bullet$-negligible. Let $a \in T_1$ and define $p_1: S \to T_1$ by

$$p_1(s) = \begin{cases} p(s) & \text{if } s \in \complement B, \\ a & \text{if } s \in B. \end{cases}$$

Then $p^{-1}(t) = p_1^{-1}(t)$ for every $t \in T_1$, $t \neq a$, $p_1: S \to T_1$ is a ν-proper mapping and $p_1(\nu) = p(\nu)_{T_1}$ (= restriction of $p(\nu)$ to T_1).

We may now state and prove the desired variant of the disintegration theorem:

Theorem 5′ (Disintegration of measures). – *Let S and T be two locally compact spaces, ν a positive Radon measure on S, $p: S \to T$ a ν-proper mapping. Let $\mu \neq 0$ be a positive Radon measure on T such that $p(\nu)$ is absolutely continuous with respect to μ and let ρ be a lifting of $M_R^\infty(T,\mu)$. Then there is a mapping $\lambda: t \to \lambda_t$ of T into $\mathcal{M}_+(S)$ appropriate with respect to (μ,ρ), such that*

$$\int_S g(f \circ p)\,d\nu = \int_T f\langle g,\lambda\rangle\,d\mu$$

for every $f \in \mathcal{K}(T)$ and $g \in \mathcal{K}(S)$. If in addition ρ is almost strong, then λ_t is concentrated on $p^{-1}(t)$, $\mu^\bullet$-almost everywhere.

Proof: Only the last statement of the theorem has to be proved since the rest is a (partial) restatement of Theorem 5.

By Remark 1) it is enough to consider the case when there is $a \in T_1$ such that $\{a\}$ *is $\mu^\bullet$-negligible.* By Remarks 3) and 4) we may assume[1]) that there is a *strong lifting* ρ_1 of $M_R^\infty(T_1,\mu_1)$ such that

$$\rho(f)|T_1 = \rho_1(f|T_1)$$

for all $f \in M_R^\infty(T,\mu)$. As in Remark 6), define $p_1: S \to T_1$ by $p_1(s) = p(s)$ for $s \in \complement B$ and $p_1(s) = a$ for $s \in B$. As in Remark 5), define $\lambda_1: T_1 \to \mathcal{M}_+(S)$ by $\lambda_1 = \lambda|T_1$.

Let $f \in \mathcal{K}(T_1)$ and let $f': T \to R$ be equal to f on T_1 and 0 on $T - T_1$. Let $g \in \mathcal{K}(S)$. By 6.2) of Proposition 6 we have

$$\int_S g(f \circ p_1)\,d\nu = \int_S g(f' \circ p)\,d\nu = \int_T f'\langle g,\lambda\rangle\,d\mu = \int_{T_1} f\langle g,\lambda_1\rangle\,d\mu_1.$$

We may now apply Theorem 5 to S, T_1, ν, p_1, μ_1 and the strong lifting ρ_1 of $M_R^\infty(T_1,\mu_1)$. Clearly λ_1 is appropriate with respect to (μ_1,ρ_1) (use Remark 5)) and λ_1 satisfies ii) of 5.1). By iii) of 5.1) there is then a set $A_1 \subset T_1$, $\mu_1^\bullet$-negligible, such that for $t \notin A_1$, λ_t is concentrated on $p_1^{-1}(t)$. Since for $t \in T_1, t \neq a, p^{-1}(t) = p_1^{-1}(t)$, and since the set $(T - T_1) \cup A_1 \cup \{a\}$ is $\mu^\bullet$-negligible, we conclude that λ_t is concentrated on $p^{-1}(t)$, $\mu^\bullet$-almost everywhere. Thus Theorem 5′ is proved.

Remarks. – 1) Theorems 5 and 5′ above generalize Theorem 1 in [15], chapter 6, pp. 58–63 (make use of Theorem 8, section 4, chapter 8), as well as Theorem 4 in [21], pp. 40–41 (see also [49] and the results in [22], section 6).

2) Theorem 2 in [15], chapter 6, pp. 64–65, for instance, can be generalized using the notion of strong lifting or almost strong lifting.

3) In connection with the disintegration of measures see also [55], [56], [60], [61], [62′], [67], [71].

---

[1]) By eventually changing ρ on a $\mu^\bullet$-negligible set.

CHAPTER X

On certain endomorphisms of $L_R^\infty(Z,\mu)$

In this chapter, unless we mention explicitly the contrary, *we shall denote by* $\mathscr{Z}_1=(Z_1,\mu_1)$ *and* $\mathscr{Z}_2=(Z_2,\mu_2)$ *two objects, where* Z_1 *and* Z_2 *are locally compact spaces,* $\mu_1 \neq 0$ *is a positive Radon measure on* Z_1 *and* $\mu_2 \neq 0$ *a positive Radon measure on* Z_2.

The main results we prove here are Theorems 1 and 2. Theorem 1 shows in particular that *every* isomorphism of $L_R^\infty(Z_1,\mu_1)$ onto $L_R^\infty(Z_2,\mu_2)$ is "induced by a point mapping" of Z_2 into Z_1. Its proof is based on the existence of a lifting. Theorem 2 gives a condition equivalent to the strong lifting property.

Theorem 1 (for Z_1, Z_2 compact spaces) and Theorem 2 were proved in [62]. Theorem 1 was generalized in [134] to the case of locally compact spaces. The method of proof presented here is essentially that of [134].

1. The spaces $\mathscr{R}(\mathscr{Z}_1,\mathscr{Z}_2)$

We shall start with the

Definition 1. – *We denote by* $\mathscr{R}(\mathscr{Z}_1,\mathscr{Z}_2)$ *the set of all representations* U *of the Banach algebra* $L_R^\infty(Z_1,\mu_1)$ *into the Banach algebra* $L_R^\infty(Z_2,\mu_2)$ *having the following properties:*

1.1) $U(\tilde{1})=\tilde{1}$.

1.2) *For every bounded set* B *of positive elements of* $L_R^\infty(Z_1,\mu_1)$, *directed for* $\leqslant$, *we have*

$$\sup U(B) = U(\sup B).$$

We note that if $U\in\mathscr{R}(\mathscr{Z}_1,\mathscr{Z}_2)$ then U is positive, that is, $U(\tilde{f})\geqslant 0$ for every $\tilde{f}\in L_R^\infty(Z_1,\mu_1)$, $\tilde{f}\geqslant 0$.

Let $B\subset L_R^\infty(Z_1,\mu_1)$ be a *bounded* set. It follows immediately from 1.1) and 1.2) that if B *is directed for* $\leqslant$, then

$$\sup U(B) = U(\sup B);$$

if *B is directed for* $\geqslant$, then

$$\inf U(B) = U(\inf B).$$

Let $\mathscr{Z}_3 = (Z_3, \mu_3)$, where Z_3 is a locally compact space and $\mu_3 \neq 0$ a positive Radon measure on Z_3. If

$$U \in \mathscr{R}(\mathscr{Z}_1, \mathscr{Z}_2) \quad \text{and} \quad V \in \mathscr{R}(\mathscr{Z}_2, \mathscr{Z}_3)$$

then

$$V \circ U \in \mathscr{R}(\mathscr{Z}_1, \mathscr{Z}_3).$$

Proposition 1. – *Let U_1 and U_2 be two elements of $\mathscr{R}(\mathscr{Z}_1, \mathscr{Z}_2)$. Then $U_1 = U_2$ if and only if $U_1(\tilde{f}) = U_2(\tilde{f})$ for all $f \in \mathscr{K}(Z_1)$.*

Proof: It is enough to show that $U_1(\tilde{f}) = U_2(\tilde{f})$ for every $f \in M_R^\infty(Z_1, \mu_1)$, $f \geqslant 0$.

Let $f \in M_R^\infty(Z_1, \mu_1)$, $f \geqslant 0$ and lower semicontinuous. Let $\mathscr{F}_0$ be the set of all functions $g \in \mathscr{K}(Z_1)$ satisfying $0 \leqslant g \leqslant f$. Then $\mathscr{F}_0$ is directed (for $\leqslant$) and (see [15], chapter 5, p. 64)

$$\tilde{f} = \sup\{\tilde{g} | g \in \mathscr{F}_0\}.$$

By 1.1) of Definition 1 we deduce $U_1(\tilde{f}) = U_2(\tilde{f})$.

Let now $f \in M_R^\infty(Z_1, \mu_1)$, $f \geqslant 0$ and let $\mathscr{F}_1$ be the set of all lower semicontinuous functions g belonging to $M_R^\infty(Z_1, \mu_1)$ and satisfying $g \geqslant f$. Then $\mathscr{F}_1$ is directed (for $\geqslant$) and (see [15], chapter 5, p. 64)

$$\tilde{f} = \inf\{\tilde{g} | g \in \mathscr{F}_1\}.$$

Whence (see the remarks following Definition 1) $U_1(\tilde{f}) = U_2(\tilde{f})$.

Hence Proposition 1 is completely proved.

2. The sets $\mathscr{U}(\mathscr{Z}_1, \mathscr{Z}_2)$ and the mappings β_u

Let $u\colon Z_2 \to Z_1$ be a mapping having the following two properties:

(1) *$f \circ u$ is $\mu_2^\bullet$-measurable for every $f \in \mathscr{K}(Z_1)$;*

(2) *$u^{-1}(K)$ is $\mu_2^\bullet$-negligible if $K \subset Z_1$ is $\mu_1^\bullet$-negligible.*

Note that if $u(Z_2)$ is relatively compact, then u belongs to the set $M_{Z_1}^\infty(Z_2, \mu_2^\bullet, \mathscr{K}(Z_2))$ (see section 5, chapter 4).

1) *Let $f \in M_R^\infty(Z_1, \mu_1)$ be a function with compact support. Then $f \circ u$ belongs to $M_R^\infty(Z_2, \mu_2)$.*

In fact, there exists a sequence (f_n) of functions belonging to $\mathscr{K}(Z_1)$ and a $\mu_1^\bullet$-negligible set $K \subset Z_1$ such that $(f_n(z))$ converges to $f(z)$ for all $z \notin K$. Then $(f_n \circ u(z))$ converges to $f \circ u(z)$ for all $z \notin u^{-1}(K)$. Since by (2), $u^{-1}(K)$ is $\mu_2^\bullet$-negligible we deduce that $f \circ u$ is $\mu_2^\bullet$-measurable and of course bounded, since f was supposed bounded. Whence

$$f \circ u \in M_R^\infty(Z_2,\mu_2).$$

If $f \in M_R^\infty(Z_2,\mu_2)$ is arbitrary then the conclusion $f \circ u \in M_R^\infty(Z_1,\mu_1)$ does not necessarily hold. For instance let $Z_2=[0,1]$ endowed with the usual topology, and μ_2 the Lebesgue measure on Z_2. Let $Z_1=[0,1]$ endowed with the discrete topology and μ_1 the Radon measure on Z_1 defined by $\mu_1(f)=\sum_{x\in Z_1} f(x)$ for $f \in \mathscr{K}(Z_1)$. Let u: $Z_2 \to Z_1$ be the identity mapping. Then u satisfies (1) and (2). However we do not necessarily have $f \circ u \in M_R^\infty(Z_2,\mu_2)$ for every $f \in M_R^\infty(Z_1,\mu_1)$.

2) *If* $f: Z_1 \to \bar{R}$ *is* $\mu_1^\bullet$*-negligible, then* $f \circ u$ *is* $\mu_2^\bullet$*-negligible.*

Hence if f and g map Z_1 into $\bar{R}$ and $f_1 \equiv f_2$, then $f_1 \circ u \equiv f_2 \circ u$.

Let $\mathscr{A}_R^\infty(Z_1,\mu_1)$ (or simply $\mathscr{A}_R^\infty$) be the subalgebra of $M_R^\infty(Z_1,\mu_1)$ consisting of all functions $f \in M_R^\infty(Z_1,\mu_1)$ having compact support and let $A_R^\infty(Z_1,\mu_1)$ (or simply A_R^∞) be its image (under the canonical mapping) in $L_R^\infty(Z_1,\mu_1)$.

If u: $Z_2 \to Z_1$ satisfies (1) and (2), we shall write

$$\beta_u(\tilde{f})=\widetilde{f \circ u} \tag{3}$$

for all $f \in \mathscr{A}_R^\infty(Z_1,\mu_1)$; then β_u is well defined as a mapping of $A_R^\infty(Z_1,\mu_1)$ into $L_R^\infty(Z_2,\mu_2)$.

We usually write $\tilde{f} \circ u$ (for $f \in \mathscr{A}_R^\infty$) instead of $\widetilde{f \circ u}$.

Clearly β_u is a *representation* of $A_R^\infty(Z_1,\mu_1)$ into $L_R^\infty(Z_1,\mu_1)$.

3) *If* D *is a bounded set of positive elements of* $L_R^\infty(Z_1,\mu_1)$ *directed for* $\leqslant$ *and if* $\sup D$ *belongs to* $A_R^\infty(Z_1,\mu_1)$ *then*

$$\beta_u(\sup D)=\sup \beta_u(D).$$

Note that our hypotheses imply that there is a compact set $K \subset Z_1$ such that $\tilde{\varphi}_K \tilde{f}=\tilde{f}$ for all $\tilde{f} \in D$.

To prove 3) we reason as follows: Let $\tilde{f}_\infty=\sup D$. Since β_u is positive we have

$$\beta_u(\tilde{f}_\infty) \geqslant \sup\{\beta_u(\tilde{f}) | \tilde{f} \in D\}.$$

Let now $(\tilde{f}_n)$ be an increasing sequence of elements of D such that $\sup_n \tilde{f}_n=\tilde{f}_\infty$. Then the sequence (f_n) converges to f, $\mu_1^\bullet$-almost everywhere, whence $(f_n \circ u)$ converges to $f_\infty \circ u$, $\mu_2^\bullet$-almost everywhere. Hence

$$\beta_u(\tilde{f}_\infty)=\tilde{f}_\infty \circ u=\sup_n \tilde{f}_n \circ u \leqslant \sup\{\beta_u(\tilde{f}) | \tilde{f} \in D\}.$$

We deduce

$$\beta_u(\sup D)=\sup \beta_u(D).$$

We shall now introduce the following basic

Definition 2. – *We denote by* $\mathscr{U}(\mathscr{Z}_1,\mathscr{Z}_2)$ *the set of all mappings* $u\colon Z_2\to Z_1$ *having the following properties:*

2.1) $f\circ u$ *is* $\mu_2^\bullet$*-measurable for every* $f\in\mathscr{K}(Z_1)$;

2.2) $u^{-1}(K)$ *is* $\mu_2^\bullet$*-negligible if* $K\subset Z_1$ *is* $\mu_1^\bullet$*-negligible;*

2.3) *If* Δ *is the set of all* $f\in\mathscr{K}(Z_1)$ *satisfying* $0\leqslant f\leqslant 1$ *then*

$$\sup_{f\in\Delta}\tilde{f}\circ u=\tilde{1}.$$

The next proposition gives several equivalent formulations for Definition 2.

Proposition 2. – *Let* $u\colon Z_2\to Z_1$ *be a mapping satisfying conditions* 2.1) *and* 2.2) *of Definition* 2. *The following assertions are equivalent:*

i) *If* Δ *is the set of all* $f\in\mathscr{K}(Z_1)$ *satisfying* $0\leqslant f\leqslant 1$, *then*

$$\sup_{f\in\Delta}\tilde{f}\circ u=\tilde{1};$$

ii) *If* Δ' *is a set of functions belonging to* $\mathscr{K}_+(Z_1)$, *directed for* $\leqslant$, *the supremum of which is* 1, *then*

$$\sup_{f\in\Delta'}\tilde{f}\circ u=\tilde{1};$$

iii) *If* $\mathscr{K}$ *is the set of all compact parts of* Z_1, *then*

$$\sup_{K\in\mathscr{K}}\tilde{\varphi}_K\circ u=\tilde{1};$$

iv) *If* $\mathscr{K}'$ *is a set of compact parts of* Z_1 *such that* $\sup\{\tilde{K}|K\in\mathscr{K}'\}=\tilde{Z}_1$, *then*

$$\sup_{K\in\mathscr{K}'}\tilde{\varphi}_K\circ u=\tilde{1}.$$

Proof: We shall show first that i) ⇒ ii). Suppose therefore that i) holds. Let $h\in\Delta$. Then $\sup_{f\in\Delta'} fh=h$ and hence (use Dini's theorem) there is an increasing sequence (f_n) of functions belonging to Δ' such that $(f_n h)$ converges uniformly to h. Then

$$\tilde{h}\circ u=\sup_n(\tilde{f}_n\tilde{h})\circ u\leqslant\sup_n\tilde{f}_n\circ u\leqslant\sup_{f\in\Delta'}\tilde{f}\circ u.$$

Since $h\in\Delta$ was arbitrary we deduce

$$\tilde{1}=\sup_{h\in\Delta}\tilde{h}\circ u\leqslant\sup_{f\in\Delta'}\tilde{f}\circ u,$$

whence

$$\sup_{f\in\Delta'}\tilde{f}\circ u=\tilde{1}.$$

Since clearly ii) ⇒ i), we deduce that i) *and* ii) *are equivalent.*

We shall now show that iii) $\Rightarrow$ iv). Suppose therefore that iii) holds. Let $L \subset Z_1$ be a compact. There is then a sequence (K_n) of compacts belonging to $\mathscr{K}'$ such that $\sup_n \tilde{K}_n \cap \tilde{L} = \tilde{L}$[1]. Hence

$$\tilde{\varphi}_L \circ u = \sup_n \tilde{\varphi}_{K_n \cap L} \circ u \leqslant \sup_n \tilde{\varphi}_{K_n} \circ u \leqslant \sup_{K \in \mathscr{K}'} \tilde{\varphi}_K \circ u.$$

Since $L \subset Z_1$ was an arbitrary compact we deduce

$$\tilde{1} = \sup_{L \in \mathscr{K}} \tilde{\varphi}_L \circ u \leqslant \sup_{K \in \mathscr{K}'} \tilde{\varphi}_K \circ u,$$

whence

$$\sup_{K \in \mathscr{K}'} \tilde{\varphi}_K \circ u = \tilde{1}.$$

Since clearly iv) $\Rightarrow$ iii), we deduce that iii) *and* iv) *are equivalent.*

To complete the proof of the proposition it is enough to prove that i) *and* iii) *are equivalent.*

To show that i) $\Rightarrow$ iii) we remark that if $f \in \Delta$ and $K = \operatorname{Supp} f$, then $K \in \mathscr{K}$ and $\tilde{f} \circ u \leqslant \tilde{\varphi}_K \circ u$. To show that iii) $\Rightarrow$ i) we remark that if $K \in \mathscr{K}$ and if $f \in \Delta$ is such that $\varphi_K \leqslant f$, then $\tilde{\varphi}_K \circ u \leqslant \tilde{f} \circ u$.

Hence i), ii), iii) and iv) are equivalent and Proposition 2 is proved.

If $\mathscr{K}'$ is a set of compacts of Z_1 then $\sup_{K \in \mathscr{K}'} \tilde{\varphi}_K \circ u = \tilde{1}$ if and only if

$$\sup \{\widetilde{u^{-1}(K)} \mid K \in \mathscr{K}'\} = \tilde{Z}_2.$$

4) *If* $u \in \mathscr{U}(\mathscr{Z}_1, \mathscr{Z}_2)$ *and* $f \in M_R^\infty(Z_1, \mu_1)$, *then the function* $f \circ u$ belongs to $M_R^\infty(Z_2, \mu_2)$.

We have, for every compact $K \subset Z_1$,

$$\varphi_{u^{-1}(K)} f \circ u = (\varphi_K \circ u)(f \circ u) = (\varphi_K f) \circ u.$$

By 1) above, the function $\varphi_{u^{-1}(K)} f \circ u$ is $\mu_2^\bullet$-measurable. Since

$$\sup \{\widetilde{u^{-1}(K)} \mid K \in \mathscr{K}\} = \tilde{Z}_2$$

we deduce (see Theorem 15, section 5, chapter 1) that $f \circ u$ is $\mu_2^\bullet$-measurable. Since $f \circ u$ is clearly bounded, we conclude $f \circ u \in M_R^\infty(Z_2, \mu_2)$.

If $u \in \mathscr{U}(\mathscr{Z}_1, \mathscr{Z}_2)$, we shall write

$$\beta_u(\tilde{f}) = \widetilde{f \circ u} \tag{4}$$

for all $f \in M_R^\infty(Z_1, \mu_1)$; then β_u is well defined as a mapping of $L_R^\infty(Z_1, \mu_1)$ into $L_R^\infty(Z_2, \mu_2)$. Note also that the mapping β_u defined by (4) on $L_R^\infty(Z_1, \mu_1)$ is an *extension* of the mapping β_u defined by (3) on $A_R^\infty(Z_1, \mu_1)$.

We usually write $\tilde{f} \circ u$ instead of $\widetilde{f \circ u}$.

---

[1]) See the remark following Theorem 14, section 5, chapter 1.

Clearly β_u is a *representation* of $L_R^\infty(Z_1,\mu_1)$ into $L_R^\infty(Z_2,\mu_2)$ and also $\beta_u(\tilde{1})=\tilde{1}$. *Hence* β_u *satisfies condition* 1.1) *of Definition* 1.

5) *Let* $u\in\mathscr{U}(\mathscr{Z}_1,\mathscr{Z}_2)$ *and let* B *be a bounded set of positive elements of* $L_R^\infty(Z_1,\mu_1)$, *directed for* $\leqslant$. *Then*

$$\beta_u(\sup B)=\sup\beta_u(B).$$

Let $\tilde{f}_\infty=\sup B$. Since β_u is positive we have

$$\beta_u(\tilde{f}_\infty)\geqslant\sup\beta_u(B);$$

conversely[1])

$$\begin{aligned}\beta_u(\tilde{f}_\infty)&=\sup_{h\in\Delta}\beta_u(\tilde{h})\,\beta_u(\tilde{f}_\infty)\\&=\sup_{h\in\Delta}\beta_u(\tilde{h}\tilde{f}_\infty)=\sup_{h\in\Delta}\Big(\sup_{\tilde{f}\in B}\beta_u(\tilde{h}\tilde{f})\Big)\\&=\sup_{\tilde{f}\in B}\Big(\sup_{h\in\Delta}\beta_u(\tilde{h}\tilde{f})\Big)=\sup_{\tilde{f}\in B}\Big(\sup_{h\in\Delta}\beta_u(\tilde{h})\,\beta_u(\tilde{f})\Big)\\&=\sup_{\tilde{f}\in B}\beta_u(\tilde{f})=\sup\beta_u(B).\end{aligned}$$

Hence β_u *satisfies condition* 1.2) *of Definition* 2, also. Therefore:

Proposition 3. – *For each* $u\in\mathscr{U}(\mathscr{Z}_1,\mathscr{Z}_2)$ *the mapping* β_u *belongs to* $\mathscr{R}(\mathscr{Z}_1,\mathscr{Z}_2)$.

We shall show in the next section that $u\to\beta_u$ is a *surjection* of $\mathscr{U}(\mathscr{Z}_1,\mathscr{Z}_2)$ onto $\mathscr{R}(\mathscr{Z}_1,\mathscr{Z}_2)$.

Let $\mathscr{Z}_3=(Z_3,\mu_3)$, where Z_3 is a locally compact space and $\mu_3\neq 0$ a positive Radon measure on Z_3. If $u\in\mathscr{U}(\mathscr{Z}_1,\mathscr{Z}_2)$ and $v\in\mathscr{U}(\mathscr{Z}_2,\mathscr{Z}_3)$ then it is easy to show that $u\circ v\in\mathscr{U}(\mathscr{Z}_1,\mathscr{Z}_3)$ and

$$\beta_{u\circ v}=\beta_v\circ\beta_u. \tag{5}$$

In fact, note first that $u\circ v$ satisfies conditions 2.1) and 2.2) of Definition 2. Further

$$\sup_{f\in\Delta}\tilde{f}\circ(u\circ v)=\sup_{f\in\Delta}((\tilde{f}\circ u)\circ v)=\Big(\sup_{f\in\Delta}(\tilde{f}\circ u)\Big)\circ v=\tilde{1}\circ v=\tilde{1}.$$

Hence $u\circ v$ satisfies also 2.3) of Definition 2. Since (5) is clearly satisfied, the assertion is proved.

3. The first main theorem

We have now established the setting for the proof of:

Theorem 1. – *The mapping* $\beta\colon u\to\beta_u$ *is a surjection of* $\mathscr{U}(\mathscr{Z}_1,\mathscr{Z}_2)$ *onto* $\mathscr{R}(\mathscr{Z}_1,\mathscr{Z}_2)$.

---

[1]) Recall that Δ is the set of all $f\in\mathscr{K}(Z_1)$ satisfying $0\leqslant f\leqslant 1$.

Proof: By Proposition 3 we have $\beta_u \in \mathscr{R}(\mathscr{Z}_1, \mathscr{Z}_2)$ for all $u \in \mathscr{U}(\mathscr{Z}_1, \mathscr{Z}_2)$. We shall show now that for every $U \in \mathscr{R}(\mathscr{Z}_1, \mathscr{Z}_2)$ there is $u \in \mathscr{U}(\mathscr{Z}_1, \mathscr{Z}_2)$ such that $U = \beta_u$; the proof of this assertion is based on the existence of a lifting of $M_R^\infty(Z_2, \mu_2)$.

Denote by Z_1^∞ the compact defined as follows: $Z_1^\infty = Z_1$ if Z_1 is compact and Z_1^∞ = the one point compactification of Z_1 if Z_1 is not compact. For each $f\colon Z_1 \to R$ denote by f' the mapping of $Z_1^\infty \to R$ defined by $f'(z) = f(z)$ if $z \in Z_1$ and $f'(z) = 0$ if $z \in Z_1^\infty - Z_1$.

Let ρ be a *lifting* of $M_R^\infty(Z_2, \mu_2)$. For every $z_2 \in Z_2$ the mapping $f \to \rho(U(\widetilde{f|Z_1}))(z_2)$ is a character of $C(Z_1^\infty)$. It follows that there is one and only one $z_1 \in Z_1^\infty$ such that

$$\rho(U(\widetilde{f|Z_1}))(z_2) = f(z_1)$$

for every $f \in C(Z_1^\infty)$. Let us write $u_1(z_2) = z_1$. Then u_1 is a well defined mapping of Z_2 into Z_1^∞ and

$$f' \circ u_1(z_2) = f'(z_1) = \rho(U(\tilde{f}))(z_2)$$

for $f \in \mathscr{K}(Z_1)$, $z_2 \in Z_2$; whence

$$f' \circ u_1 = \rho(U(\tilde{f})) \tag{1}$$

for every $f \in \mathscr{K}(Z_1)$.

Let $f \in M_R^\infty(Z_1, \mu_1)$ be positive and lower semicontinuous and let $\Delta(f)$ be the set of all $g \in \mathscr{K}(Z_1)$ satisfying $0 \leqslant g \leqslant f$. Then (use (1))

$$f' \circ u_1 = \sup_{g \in \Delta(f)} g' \circ u_1 = \sup_{g \in \Delta(f)} \rho(U(\tilde{g})) \leqslant \rho(U(\tilde{f}));$$

we conclude (use Theorem 3, section 3, chapter 3 and 1.2) of Definition 1) that $f' \circ u_1 \equiv \rho(U(\tilde{f}))$.

If $f = 1$ we obtain[1])

$$\varphi_{u_1^{-1}(Z_1)} = \varphi_{Z_1} \circ u_1 \equiv \rho(U(\tilde{1})) = 1;$$

whence $Z_2 - u_1^{-1}(Z_1)$ is $\mu_2^\bullet$-negligible.

Let now $\omega \in Z_1$ be fixed and let $u\colon Z_2 \to Z_1$ be defined by $u(z_2) = u_1(z_2)$ if $z_2 \in u_1^{-1}(Z_1)$ and $u(z_2) = \omega$ if $z_2 \in Z_2 - u_1^{-1}(Z_1)$. Clearly $f \circ u \equiv f' \circ u_1$ for every function $f\colon Z_1 \to R$.

From (1) we deduce that *$f \circ u$ is $\mu_2^\bullet$-measurable for every $f \in \mathscr{K}(Z_1)$*. Let now $K \subset Z_1$ be $\mu_1^\bullet$-negligible and let $\nabla(\varphi_K)$ be the set of all functions $g \in M_R^\infty(Z_1, \mu_1)$ which are lower semicontinuous and superior to φ_K. Then $\inf_{g \in \nabla(\varphi_K)} \tilde{g} = 0$ (see [15], chapter 5, p. 64), whence $\inf_{g \in \nabla(\varphi_K)} U(\tilde{g}) = 0$. We deduce (use Theorem 3, section 3, chapter 3)

$$\inf_{g \in \nabla(\varphi_K)} \rho(U(\tilde{g})) \equiv 0.$$

---

[1]) The characteristic function φ_{Z_1} written below is defined on Z_1^∞.

Since $\varphi'_K \circ u_1 \leqslant g' \circ u_1 \leqslant \rho(U(\tilde{g}))$ for all $g \in \nabla(\varphi_K)$ we deduce $\varphi'_K \circ u_1 \equiv 0$, and thus $\varphi_K \circ u \equiv 0$; hence $u^{-1}(K)$ is $\mu_2^\bullet$-negligible.

Therefore $u: Z_2 \to Z_1$ *satisfies the conditions* 2.1) *and* 2.2) *of Definition* 2. We may then define the mapping β_u (see (3) of section 2) on $A_R^\infty(Z_1, \mu_1)$. By (1)

$$\beta_u(\tilde{f}) = U(\tilde{f})$$

for all $f \in \mathscr{K}(Z_1)$. Using 1.2) of Definition 1 we deduce that u *satisfies* 2.3) *of Definition* 2 also; whence $u \in \mathscr{U}(\mathscr{Z}_2, \mathscr{Z}_1)$. Since

$$\beta_u(\tilde{f}) = U(\tilde{f})$$

for all $f \in \mathscr{K}(Z_1)$, we deduce (use Propositions 3 and 1) that $\beta_u = U$.

Hence Theorem 1 is completely proved.

Remarks. – Let u be the mapping constructed in Theorem 1. Using (1) we deduce that there exists a $\mu_2^\bullet$-negligible set $A_1 \subset Z_2$ such that

$$\rho(f \circ u)(t) = f \circ u(t)$$

for all $f \in \mathscr{K}(Z_1)$ and $t \notin A_1$. Let now $v \in \mathscr{U}(\mathscr{Z}_1, \mathscr{Z}_2)$ be a second mapping satisfying $\beta_v = U$. Suppose that there is a $\mu_2^\bullet$-negligible set $A_2 \subset Z_2$ such that

$$\rho(f \circ v)(t) = f \circ v(t)$$

for all $f \in \mathscr{K}(Z_1)$ and $t \notin A_2$. Then

$$f \circ u(t) = f \circ v(t) (= \rho(U(\tilde{f}))(t))$$

for all $f \in \mathscr{K}(Z_1)$ and $t \notin A_1 \cup A_2$. Hence *u and v coincide $\mu_2^\bullet$-almost everywhere.*

If Z_1 is compact then $Z_1^\infty = Z_1$ and $u_1 = u$. It follows that

$$\rho(f \circ u) = f \circ u$$

for all $f \in \mathscr{K}(Z_1)$. Hence (with the notations of chapter 4, section 5) $\rho(u) = u$. If $v \in \mathscr{U}(\mathscr{Z}_1, \mathscr{Z}_2)$ is a second mapping such that $\rho(v) = v$ and $U = \beta_v$ we deduce $u = v$. Hence *if Z_1 is compact and the lifting ρ is given*, then there is a unique mapping $u \in \mathscr{U}(\mathscr{Z}_1, \mathscr{Z}_2)$ satisfying

$$U = \beta_u \quad \text{and} \quad \rho(u) = u.$$

The examples and results that follow in this and in the next section give supplementary information concerning the connection between $U(= \beta_u)$ and u.

Example 1. – Let T^1 be the group $\{t \text{ complex} \mid |t| = 1\}$ and let Q_0 be the set of all $t = e^{2\pi\theta i} \neq 1$, with θ rational. Let now $Z = (T^1)^R$ and let μ be the normalized Haar measure on Z (recall that $\mu = \bigotimes_{i \in R} \mu_i$, where μ_i is the normalized Haar measure on T^1). Define $\mathscr{Z}$ to be (Z, μ)

Let $\alpha: T^1 \to T^1$ be the mapping defined by

$$\alpha(z) = \begin{cases} z & \text{if } z \notin Q_0, \\ 1 & \text{if } z \in Q_0. \end{cases}$$

Let $u_1: Z \to Z$ be the *identity mapping* and let $u_2: Z \to Z$ be the mapping defined by

$$u_2((x_t)_{t\in R}) = (\alpha(x_t))_{t\in R}.$$

Before proceeding further, let us establish that

$$f \circ u_2 \equiv f \tag{2}$$

for all $f \in C(Z)$. By the Stone-Weierstrass theorem it is enough to verify this for functions f depending on one coordinate only. If f depends on t_0, then $f = f_1 \circ pr_{t_0}$ where $f_1 \in C(T^1)$. Then

$$\begin{aligned} \{z | f(u_2(z)) \neq f(z)\} &= \{z | f_1(pr_{t_0}(u_2(z))) \neq f_1(pr_{t_0}(z))\} \\ &= \{(x_t)_{t\in R} | f_1(\alpha(x_{t_0})) \neq f_1(x_{t_0})\} \\ &\subset \{(x_t)_{t\in R} | \alpha(x_{t_0}) \neq x_{t_0}\} \\ &\subset \{(x_t)_{t\in R} | x_{t_0} \in Q_0\}; \end{aligned}$$

since this last set is $\mu^\bullet$-negligible we deduce that $f\ u_2 \equiv f$. Hence (2) is proved.

Note now that $u_2 \in \mathscr{U}(\mathscr{Z}, \mathscr{Z})$. By (2), u_2 satisfies condition 2.1) of Definition 2. To verify 2.2) we reason as follows: By (2) we have

$$\int_Z f \circ u_2 \, d\mu = \int_Z f \, d\mu \tag{3}$$

for all $f \in C(Z)$, hence for all bounded *Baire* functions $f: Z \to R$. Now let $K \subset Z$ be $\mu^\bullet$-negligible. Since μ is "completion regular" (see [50], p. 287) there exists a $\mu^\bullet$-negligible Baire set $K' \supset K$. Then $u_2^{-1}(K')$ is $\mu^\bullet$-negligible, whence $u_2^{-1}(K)$ is $\mu^\bullet$-negligible. We deduce that u_2 satisfies 2.2) and hence (since Z is compact) that $u_2 \in \mathscr{U}(\mathscr{Z}, \mathscr{Z})$.

From (2) we deduce also that

$$\beta_{u_1} = \beta_{u_2}$$

(= the identity mapping of $L_R^\infty(Z, \mu)$ onto $L_R^\infty(Z, \mu)$).

We shall now show that if

$$A = \{z | u_1(z) \neq u_2(z)\} \quad \textit{then} \quad \mu^\bullet(A) = 1. \tag{4}$$

For this we first note that $Z - u_2(Z) \subset A$. In fact, if $z \in Z - u_2(Z)$, then $z \neq u_2(z)$; whence $u_1(z) \neq u_2(z)$ and hence $z \in A$.

We now show that

$$\mu^\bullet(Z - u_2(Z)) = 1. \tag{5}$$

If the inequality $\mu^\bullet(Z-u_2(Z))<1$ were satisfied, then there would be a Baire set $B'\supset Z-u_2(Z)$ satisfying $\mu^\bullet(B')<1$. Since B' is a Baire set, there is a countable set $I\subset R$ and a set $B''\subset(T^1)^I$ such that

$$\{(x_t)_{t\in R}|(x_t)_{t\in I}\in B''\}=B'.$$

Since $B'\neq Z$, clearly $B''\neq(T^1)^I$. Now let $y_0=(x_t)_{t\in R}$ be such that $(x_t)_{t\in I}\notin B''$ and $x_{t_0}\in Q_0$ for some $t_0\notin I$. Then $y_0\notin B'$ and $y_0\notin u_2(Z)$; whence $y_0\in Z-u_2(Z)$. Since this leads to a contradiction, we have established (5).

If the set $u_2(Z)$ were $\mu^\bullet$-negligible then (since $u_2\in\mathscr{U}(\mathscr{Z},\mathscr{Z})$) we would deduce that $Z=u_2^{-1}(u_2(Z))$ is also $\mu^\bullet$-negligible. Hence $u_2(Z)$ is not $\mu^\bullet$-negligible and hence $u_2(Z)$ is *not* $\mu^\bullet$-measurable.

Therefore:

(6) $U=\beta_{u_1}=\beta_{u_2}$ *does not necessarily imply that* u_1 *and* u_2 *coincide* $\mu^\bullet$*-almost everywhere*, even if U is an isomorphism of $L_R^\infty(Z,\mu)$ onto $L_R^\infty(Z,\mu)$.

(7) $U=\beta_u$ *does not necessarily imply that* $u(Z)$ *is* $\mu^\bullet$*-measurable or that* $Z-u(Z)$ *is* $\mu^\bullet$*-negligible*, even if U is an isomorphism of $L_R^\infty(Z,\mu)$ onto $L_R^\infty(Z,\mu)$.

Example 2. – Let $\mathscr{Z}_1=(X,\mu)$ and $\mathscr{Z}_2=(B,\beta_B)$ be as in the Example of section 3, chapter 8. Then $U:\tilde{f}\to\widetilde{f|B}$ is an isomorphism of $L_R^\infty(X,\mu)$ onto $L_R^\infty(B,\beta_B)$. Let $V=U^{-1}$; whence V is the isomorphism of $L_R^\infty(B,\beta_B)$ onto $L_R^\infty(X,\mu)$, inverse to U. Let $h\in\mathscr{U}(\mathscr{Z}_2,\mathscr{Z}_1)$ $(h: X\to B)$ be such that $V=\beta_h$. Since B is metrizable, h is $\mu^\bullet$-measurable (see section 4 in this chapter). It follows then that h is *not* almost injective[1]).

In the next proposition we discuss some of the connections between U $(=\beta_u)$ and u, when U is *injective*.

Proposition 4. – *Let* $U\in\mathscr{R}(\mathscr{Z}_1,\mathscr{Z}_2)$ *and* $u\in\mathscr{U}(\mathscr{Z}_1,\mathscr{Z}_2)$ *such that* $\beta_u=U$. *The following are then equivalent:*

h) U *is injective;*

hh) *a set* $K\subset Z_1$ *which is* $\mu_1^\bullet$*-measurable is* $\mu_1^\bullet$*-negligible if and only if* $u^{-1}(K)$ *is* $\mu_2^\bullet$*-negligible.*

Moreover if U *is injective then*[2]) $(\mu_1)_\bullet(Z_1-u(Z_2))=0$; *the converse is not true.*

---

[1]) We say that a mapping $f: Z_1\to Z_2$ is almost injective if there is a $\mu_1^\bullet$-negligible set $A\subset Z_1$ such that $f|(Z_1-A)$ is injective. To prove the assertion that h is not almost injective use approximation by functions in $C(X)$.

[2]) If Z is a locally compact space, μ a positive Radon measure on Z and $A\subset Z$, then we define

$$\mu_\bullet(A)=\sup\{\mu(K)|K\text{ compact}\subset A\}.$$

Proof: h)$\Rightarrow$hh). We already know that if $K \subset Z_1$ is $\mu_1^\bullet$-negligible then $u^{-1}(K)$ is $\mu_2^\bullet$-negligible. Let now $K \subset Z_1$ be $\mu_1^\bullet$-measurable. Suppose that $u^{-1}(K)$ is $\mu_2^\bullet$-negligible; then

$$U(\tilde{\varphi}_K) = \beta_u(\tilde{\varphi}_K) = \tilde{\varphi}_K \circ u = \tilde{\varphi}_{u^{-1}(K)} = 0$$

and hence, since U is injective, $\tilde{\varphi}_K = 0$. Hence $\varphi_K \equiv 0$ and hence K is $\mu_1^\bullet$-negligible.

hh)$\Rightarrow$h). Suppose now that hh) is satisfied. Let $\tilde{f} \in L_R^\infty(Z_1, \mu_1)$ be such that $U(\tilde{f}) = 0$. Then $U(\tilde{g}) = U(\tilde{f})U(\tilde{f}) = 0$ if $g = f^2$. Let now $\lambda > 0$ and $K \subset Z_1$ a compact such that $g \geqslant \lambda \varphi_K$. Then $U(\tilde{g}) \geqslant \lambda U(\tilde{\varphi}_K)$, that is, $U(\tilde{\varphi}_K) = 0$. Hence $\beta_u(\tilde{\varphi}_K) = \tilde{\varphi}_K \circ u = 0$, that is, $\varphi_K \circ u(x) = \varphi_{u^{-1}(K)}(x) = 0$, $\mu_2^\bullet$-almost everywhere. Hence $u^{-1}(K)$ is $\mu_2^\bullet$-negligible and hence K is $\mu_1^\bullet$-negligible. Since K was arbitrary we deduce $\tilde{g} = 0 \Rightarrow \tilde{f} = 0 \Rightarrow U$ is injective.

Suppose now U *injective;* then

(8) $$(\mu_1)_\bullet(Z_1 - u(Z_2)) = 0.$$

In fact, let $K \subset Z_1 - u(Z_2)$ be a compact set. Then $\varphi_K \circ u = 0 \Rightarrow \beta_u(\tilde{\varphi}_K) = 0 \Rightarrow U(\tilde{\varphi}_K) = 0 \Rightarrow \tilde{\varphi}_K = 0$ (since U is injective). Hence (8) is proved.

Let us show now that the converse is not true; we mean by this that we may have $(\mu_1)_\bullet(Z_1 - u(Z_2)) = 0$ *without* U being injective. In fact: Let $A \subset (0,1)$ and $B \subset (0,1)$ be two compact sets such that

(9) $l(A) = 0$ (l = the Lebesgue measure on R);

(10) $$l(B) > 0;$$

(11) There is a continuous bijection φ of A onto B.

Let now $Z_1 = [-1,0] \cup B$, $Z_2 = [-1,0] \cup A$, $\mu_1 = l_{Z_1}$ and $\mu_2 = l_{Z_2}$. Let $u: Z_2 \to Z_1$ be defined by

$$u(z_2) = \begin{cases} z_2 & \text{if } z_2 \in [-1,0], \\ \varphi(z_2) & \text{if } z_2 \in A. \end{cases}$$

Then u is a continuous bijection of Z_2 onto Z_1. Hence if $f \in C(Z_1)$ then $f \circ u \in C(Z_2)$ and hence $f \circ u$ is $\mu_2^\bullet$-measurable. Let now $K \subset Z_1$ and note that $u^{-1}(K) \subset (K \cap [-1,0]) \cup A$. We deduce that $u^{-1}(K)$ is $\mu_2^\bullet$-negligible if K is $\mu_1^\bullet$-negligible. Hence $u \in \mathcal{U}(\mathcal{Z}_1, \mathcal{Z}_2)$.

Remark that in this case $Z_1 - u(Z_2) = \emptyset$. However β_u *is not injective.* We may verify this as follows: We have $\tilde{\varphi}_B \in L_R^\infty(Z_1, \mu_1)$ and $\tilde{\varphi}_B \neq 0$; however

$$\beta_u(\tilde{\varphi}_B) = \tilde{\varphi}_B \circ u = \tilde{\varphi}_{u^{-1}(B)} = \tilde{\varphi}_A = 0.$$

Remark. – Suppose that U is injective. From Proposition 4 we deduce that if $u(Z_2)$ is $\mu_1^{\bullet}$-measurable then:

$$Z_1 - u(Z_2) \quad \text{is} \quad \mu_1^{\bullet}\text{-negligible.}$$

Note that if u is continuous then $u(Z_2)$ is compact, whence $\mu_1^{\bullet}$-measurable. If in particular

$$\operatorname{Supp}\mu_1 = Z_1$$

and if u is continuous *then* $u(Z_2)=Z_1$.

We have seen in Example 1 (see (7)) that $u(Z_2)$ is not necessarily $\mu_1^{\bullet}$-measurable, even if U is an isomorphism.

4. The spaces $\mathscr{U}^*(\mathscr{Z}_1,\mathscr{Z}_2)$

In this section we continue to discuss and study the connection between $U=\beta_u$ and u, under supplementary conditions. We first introduce the following:

Definition 3. – *We denote by $\mathscr{U}^*(\mathscr{Z}_1,\mathscr{Z}_2)$ the set of all mappings $u: Z_2 \to Z_1$ having the following properties:*

3.1) *u is $\mu_2^{\bullet}$-measurable;*

3.2) *$u^{-1}(K)$ is $\mu_2^{\bullet}$-negligible if $K \subset Z_1$ is $\mu_1^{\bullet}$-negligible.*

Note that if u is $\mu_2^{\bullet}$-measurable then $f\circ u$ is $\mu_2^{\bullet}$-measurable for every $f \in \mathscr{K}(Z_1)$ (see [15], chapter 4, p. 174). Hence if $u \in \mathscr{U}^*(\mathscr{Z}_1,\mathscr{Z}_2)$ then u has the properties 2.1) and 2.2) of Definition 2. Further if u is $\mu_2^{\bullet}$-measurable then the set of all compact parts $L \subset Z_2$ such that $u|L$ is continuous is μ_2-dense (see [15], chapter 4, p. 192). We deduce that iii) of Proposition 2 is satisfied. Hence if $u \in \mathscr{U}^*(\mathscr{Z}_1,\mathscr{Z}_2)$ then u has property 2.3) of Definition 2. Thus:

$$\mathscr{U}^*(\mathscr{Z}_1,\mathscr{Z}_2) \subset \mathscr{U}(\mathscr{Z}_1,\mathscr{Z}_2).$$

1) If Z_1 *is metrizable then* $\mathscr{U}(\mathscr{Z}_1,\mathscr{Z}_2)=\mathscr{U}^*(\mathscr{Z}_1,\mathscr{Z}_2)$.

This follows from 2.3) of Definition 2, statement 4) of section 2 in this chapter and Theorem 4, [15], chapter 4, p. 179.

Let $\mathscr{Z}_3=(Z_3,\mu_3)$, where Z_3 is a locally compact space and $\mu_3 \neq 0$ a positive Radon measure on Z_3.

2) *If* $u \in \mathscr{U}^*(\mathscr{Z}_1,\mathscr{Z}_2)$ and $v \in \mathscr{U}^*(\mathscr{Z}_2,\mathscr{Z}_3)$ then $u\circ v \in \mathscr{U}^*(\mathscr{Z}_1,\mathscr{Z}_3)$.

We know already that $u\circ v \in \mathscr{U}(\mathscr{Z}_1,\mathscr{Z}_3)$. Hence it remains to show that $u\circ v$ is $\mu_3^{\bullet}$-measurable. For this we shall reason as follows: Let $K \subset Z_3$ be a compact. There exists then a sequence (K_n) of compact sets and a $\mu_3^{\bullet}$-negligible set N_3 such that $N_3 \cup \bigcup_n K_n = K$ and $v|K_n$ is continuous for all n. Let now n be fixed and consider the compact $v(K_n)$. Since u is $\mu_2^{\bullet}$-measurable there exists a sequence $(L_p^{(n)})$ of compacts and a $\mu_2^{\bullet}$-negligible set $N_2^{(n)}$ such that $N_2^{(n)} \cup \bigcup_p L_p^{(n)} = v(K_n)$ and $u|L_p^{(n)}$ is continuous for all p. Let $N_1^{(n)} = v^{-1}(N_2^{(n)}) \cap K_n$ and for each p let $C_{p,n} = v^{-1}(L_p^{(n)}) \cap K_n$. Clearly $C_{p,n}$ is compact for each p and $K_n = N_1^{(n)} \cup \left(\bigcup_p C_{p,n}\right)$. Let now $N = N_3 \cup \left(\bigcup_n N_1^{(n)}\right)$ and consider the double sequence $(C_{p,n})$. Then N is $\mu_3^\bullet$-negligible, $K = N \cup \bigcup_{(p,n)} C_{p,n}$ and $u \circ v | C_{p,n}$ is continuous for all (n,p). Since $K \subset Z_3$ was an arbitrary compact, the function $u \circ v$ is $\mu_3^\bullet$-measurable.

Recall that if u_1 and u_2 are two functions belonging to $\mathscr{U}^*(\mathscr{Z}_1, \mathscr{Z}_2)$ then $\beta_{u_1} = \beta_{u_2}$ if and only if $f \circ u_1 \equiv f \circ u_2$ for all $f \in \mathscr{K}(Z_1)$.

3) *Let u_1 and u_2 be two functions belonging to $\mathscr{U}^*(\mathscr{Z}_1, \mathscr{Z}_2)$ then $\beta_{u_1} = \beta_{u_2}$ if and only if the set*

$$\{z_2 | u_1(z_2) \neq u_2(z_2)\}$$

is $\mu_2^\bullet$-negligible.

To prove this assertion it is enough to consider a family $(K_j)_{j \in J}$ belonging to $\mathscr{C}(Z_2, \mu_2)$ such that:

a) $\mathrm{Supp}(\mu_2)_{K_n} = K_n$ for all n;
b) $u_1 | K_n$ and $u_2 | K_n$ are continuous for each n.

4) *Suppose Z_2 countable at infinity. Let $u \in \mathscr{U}^*(\mathscr{Z}_1, \mathscr{Z}_2)$ and suppose that β_u is injective. If μ_2 is concentrated on C, then μ_1 is concentrated on $u(C)$. In particular, the set $u(Z_2)$ is $\mu_1^\bullet$-measurable and*

$$\mu_1^\bullet(Z_1 - u(Z_2)) = 0.$$

Let (K_n) be a sequence of compact parts of C such that $C - \bigcup_n K_n$ is $\mu_2^\bullet$-negligible and such that $u | K_n$ is continuous for all n. Let $K_\infty = \bigcup_n u(K_n)$. Then K_∞ is $\mu_1^\bullet$-measurable and $K_\infty \subset u(C)$. Moreover

$$u^{-1}(Z_1 - K_\infty) \subset Z_2 - \bigcup_n K_n$$

and hence $u^{-1}(Z_1 - K_\infty)$ is $\mu_2^\bullet$-negligible. By Proposition 4, $Z_1 - K_\infty$ is $\mu_1^\bullet$-negligible. Since $Z_1 - u(C)$ is contained in $Z_1 - K_\infty$, μ_1 is concentrated on $u(C)$.

Proposition 5. – *Suppose Z_1 and Z_2 countable at infinity and let $u \in \mathscr{U}^*(\mathscr{Z}_1, \mathscr{Z}_2)$ and $v \in \mathscr{U}^*(\mathscr{Z}_2, \mathscr{Z}_1)$. Suppose that:*

(1) $$\beta_u \circ \beta_v = I \quad \text{and} \quad \beta_v \circ \beta_u = I.$$

Then there are, a $\mu_1^\bullet$-negligible set $N_1 \subset Z_1$ and a $\mu_2^\bullet$-negligible set $N_2 \subset Z_2$ such that:

(2) $$u(Z_2 - N_2) = Z_1 - N_1 \quad \text{and} \quad v(Z_1 - N_1) = Z_2 - N_2;$$

(3) $u(v(z_1)) = z_1$ *for* $z_1 \in Z_1 - N_1$ *and* $v(u(z_2)) = z_2$ *for* $z_2 \in Z_2 - N_2$.

Proof: From (1) we deduce $\beta_{u \circ v} = I$ and $\beta_{v \circ u} = I$. It follows from 3) that there are *negligible* sets $L_1 \subset Z_1$ and $L_2 \subset Z_2$ such that:

$$u(v(z_1)) = z_1 \quad \text{if} \quad z_1 \notin L_1;$$
$$v(u(z_2)) = z_2 \quad \text{if} \quad z_2 \notin L_2.$$

Let now

$$A = u(v(Z_1 - L_1) - L_2) \quad \textit{and} \quad B = v(A).$$

Then μ_1 is concentrated on $Z_1 - L_1$ and hence μ_2 is concentrated on $v(Z_1 - L_1)$. Since L_2 is $\mu_2^\bullet$-negligible we deduce that μ_2 is concentrated on $v(Z_1 - L_1) - L_2$. We conclude that μ_1 *is concentrated on* A. Since $B = v(A)$ we also conclude that μ_2 *is concentrated on* B.

Let now

$$N_1 = Z_1 - A \quad \text{and} \quad N_2 = Z_2 - B.$$

Then $N_1 \subset Z_1$ and $N_2 \subset Z_2$ are negligible.

Clearly

$$v(Z_1 - N_1) = v(A) = B = Z_2 - N_2.$$

Let now $z_1 \in A$; then $z_1 = u(z_2)$ with $z_2 \notin L_2$. Hence $u(v(z_1)) = u(v(u(z_2)))$ $= u(z_2) = z_1$, that is,

$$u(v(z_1)) = z_1 \quad \textit{if} \quad z_1 \in A.$$

We deduce that $u(v(A)) = A$, whence $u(B) = A$. Clearly then

$$u(Z_2 - N_2) = u(B) = A = Z_1 - N_1.$$

Hence (2) *is proved*. Note also that if $z_1 \in Z_1 - N_1 \Rightarrow z_1 \in A$, whence $u(v(z_1)) = z_1$; hence the first assertion in (3) is proved. If $z_2 \in Z_2 - N_2$ then $z_2 \in B = v(A) \Rightarrow z_2 = v(z_1)$ with $z_1 \in A \Rightarrow v(u(z_2)) = v(u(v(z_1))) = v(z_1) = z_2$. Hence (3) is completely proved. Thus the proposition is proved.

Proposition 5 (see also assertion 1) above) shows that if Z_1 and Z_2 are in addition *metrizable* and if $U: L_R^\infty(Z_1,\mu_1) \to L_R^\infty(Z_2,\mu_2)$ is an isomorphism "induced" by $u \in \mathscr{U}(\mathscr{Z}_1, \mathscr{Z}_2) = \mathscr{U}^*(\mathscr{Z}_1, \mathscr{Z}_2)$, then U^{-1} is induced by "the mapping inverse to the mapping u".

Theorem 1 combined with this remark yields then (in the setting of Radon measures) several classical results (see [28], p. 335–338, [103], p. 574–586, [110], p. 83–102; see also [22], p. 84–85 and [90], p. 702–707).

5. A condition equivalent with the strong lifting property

Suppose now that Z_1 is *compact*, $\operatorname{Supp} \mu_1 = Z_1$ and that $\mathscr{Z}_2 = (Z_2, \mu_2)$ is the *hyperstonean space* associated to $L_R^\infty(Z_1,\mu_1)$. Let U be an isomorphism of $L_R^\infty(Z_1,\mu_1)$ onto $L_R^\infty(Z_2,\mu_2)$. Each class $\tilde{f} \in L_R^\infty(Z_2,\mu_2)$ contains one and only one continuous function f^*. Denote by ρ'' the lifting of $L_R^\infty(Z_2,\mu_2)$ defined by the equations:

$$\rho''(f) = f^* \quad \text{for every} \quad f \in M_R^\infty(Z_2,\mu_2).$$

By the Remarks following Theorem 1, there is a *unique* $u \in \mathscr{U}(\mathscr{Z}_1, \mathscr{Z}_2)$ such that

$$\beta_u = U \quad \text{and} \quad f \circ u = \rho''(U(\tilde{f})) \quad \textit{for all} \quad f \in C(Z_1).$$

Since $f \circ u$ is continuous for all $f \in C(Z_1)$, we deduce that $u: Z_2 \to Z_1$ is continuous. Using the Remark following Proposition 4, we conclude that the mapping u is *surjective.*

We shall say that a mapping $\psi: Z_1 \to Z_2$ satisfies condition hh) if (see the statement of Proposition 4):

A set $K \subset Z_2$ which is $\mu_2^\bullet$-measurable is $\mu_2^\bullet$-negligible if and only if $\psi^{-1}(K)$ is $\mu_1^\bullet$-negligible.

Theorem 2. – *The following conditions concerning (Z_1, μ_1) are equivalent:*

a) *There is a strong lifting of $M_R^\infty(Z_1, \mu_1)$.*

b) *There is an injection $v \in \mathscr{U}(\mathscr{Z}_2, \mathscr{Z}_1)$ satisfying condition* hh) *and such that $u(v(z_1)) = z_1$ for every $z_1 \in Z_1$.*

Proof: a)$\Rightarrow$b). Let ρ' be a strong lifting of $M_R^\infty(Z_1, \mu_1)$. By Theorem 1 (see also the Remarks following the proof) there is $v \in \mathscr{U}(\mathscr{Z}_2, \mathscr{Z}_1)$ such that

$$\beta_v = U^{-1} \quad \text{and} \quad g \circ v = \rho'(U^{-1}(\tilde{g})) \quad \text{for} \quad g \in C(Z_2).$$

If $f \in C(Z_1)$ we have

$$f \circ (u \circ v) = (f \circ u) \circ v = \rho'(U^{-1}(\tilde{f} \circ u))$$

since $f \circ u \in C(Z_2)$. But

$$U^{-1}(\tilde{f} \circ u) = U^{-1}(U(\tilde{f})) = \tilde{f}$$

and hence

$$f \circ (u \circ v) = \rho'(\tilde{f}) = f$$

(since ρ' is strong). Since $f \in C(Z_1)$ was arbitrary we deduce $u(v(z_1)) = z_1$ for every $z_1 \in Z_1$. From these relations it follows immediately that v is *injective.* Moreover U^{-1} is an isomorphism (whence injective). By Proposition 4, the function v satisfies hh).

Hence we proved that a)$\Rightarrow$b).

b)$\Rightarrow$a). Define ρ' by the equations

$$\rho'(f) = \rho''(U(\tilde{f})) \circ v \tag{1}$$

for $f \in M_R^\infty(Z_1, \mu_1)$. It is easy to see that ρ' is a lifting of $M_R^\infty(Z_1, \mu_1)$ (see the Remark at the end of the proof). If $f \in C(Z_1)$ we have

$$\rho'(f) = \rho''(U(\tilde{f})) \circ v = (f \circ u) \circ v = f \circ (u \circ v) = f.$$

Hence ρ' is strong.

Hence we have proved that b)$\Rightarrow$a). Therefore Theorem 2 is completely proved.

Remark. – Note that if v satisfies b) then $\beta_v = U^{-1}$. In fact, since $u \circ v =$ the identity mapping of Z_1 we have

$$\beta_v(U(\tilde{f})) = (U(\tilde{f})) \circ v = (\tilde{f} \circ u) \circ v = \tilde{f} \circ (u \circ v) = \tilde{f}.$$

Since U is a bijection we deduce that $\beta_v = U^{-1}$.

Appendix I. Some ergodic theorems

The spaces $\mathscr{L}^p$ $(1\leqslant p\leqslant +\infty)$ and L^p $(1\leqslant p\leqslant +\infty)$ correspond to a system $(X,N,\mathscr{R})$ (see chapter 1). We assume here that $N=\bar{N}$.

We recall (see section 3, chapter 2) that a *Dunford-Schwartz operator* is a linear mapping $T: L^1\cap L^\infty\to L^1\cap L^\infty$ such that[1] $\|T\|_1\leqslant 1$ and $\|T\|_\infty\leqslant 1$ and that we denoted by $\mathscr{D}$ the set of all Dunford-Schwartz operators. If $T\in\mathscr{D}$, then by the Riesz-Thorin convexity theorem (see [137], p. 95–96) $\|T\|_p\leqslant 1$ for each $1\leqslant p<+\infty$ and hence T can be extended by continuity to L^p (we denote the extension by the same letter). For $T\in\mathscr{D}$ and $f\in\mathscr{V}=\bigcup_{1\leqslant p<+\infty}\mathscr{L}^p$ we denote by Tf a (determined) representative of the class $T\tilde{f}$.

We shall first prove a maximal ergodic theorem[2] for a finite sequence of Dunford-Schwartz operators satisfying certain algebraic relations.

For each $f\in\mathscr{V}$ and $a>0$ we write

$$E_f(a)=\{x\,|\,|f(x)|>a\}.$$

Consider now $n+1$ *Dunford-Schwartz operators* $T_0, T_1,\ldots, T_n$ $(n\geqslant 1)$. *Let* $f\in\mathscr{L}^p\subset\mathscr{V}(1\leqslant p<+\infty)$ *and let* $a>0$. We define d_0 and f_0 by the equations:

$$\begin{aligned} d_0(x)&=af(x)/|f(x)| \quad &&\text{if}\quad a<|f(x)|,\\ d_0(x)&=f(x) &&\text{if}\quad a\geqslant|f(x)|,\\ f_0&=f-d_0. \end{aligned}$$

Since $\mu(E_f(a))<+\infty$ it follows that $d_0\in\mathscr{L}^p$; hence $f_0\in\mathscr{L}^p$. It is also obvious that $a-|d_0(x)|\geqslant 0$ for all $x\in X$. Suppose now that

---

[1]) We recall that for each $1\leqslant p\leqslant +\infty$

$$\|T\|_p=\sup\{N_p(T\tilde{f})\,|\,\tilde{f}\in L^1\cap L^\infty,\ N_p(\tilde{f})\leqslant 1\}.$$

[2]) See [68].

$0 \leqslant k \leqslant n-1$, that $d_0, \ldots, d_k, f_0, \ldots, f_k$ were defined, belong to $\mathscr{L}^p$ and

$$D_k(x) = a - \sum_{j=0}^{k} |d_j(x)| \geqslant 0$$

for all $x \in X$. We define d_{k+1} and f_{k+1} by the equations:

$$\begin{aligned} d_{k+1}(x) &= D_k(x)\, T_{k+1} f_k(x) / |T_{k+1} f_k(x)| \quad \text{if} \quad D_k(x) < |T_{k+1} f_k(x)|, \\ d_{k+1}(x) &= T_{k+1} f_k(x) \quad \text{if} \quad D_k(x) \geqslant |T_{k+1} f_k(x)|, \\ f_{k+1} &= T_{k+1} f_k - d_{k+1}. \end{aligned}$$

It is easy to see that d_{k+1} and f_{k+1} belong to $\mathscr{L}^p$ and that

$$D_{k+1}(x) = a - \sum_{j=0}^{k+1} |d_j(x)| \geqslant 0$$

for all $x \in X$. By induction we then define the functions $d_0, \ldots, d_n$, $f_0, \ldots, f_n$. Without difficulty we see that:

(1) $d_0, \ldots, d_n, f_0, \ldots, f_n$ belong to $\mathscr{L}^p$;

(2) $\sum_{j=0}^{n} |d_j(x)| \leqslant a$ for all $x \in X$;

(3) $|d_0(x)| + |f_0(x)| = |f(x)|$ for all $x \in X$;

(4) $|d_{k+1}(x)| + |f_{k+1}(x)| = |T_{k+1} f_k(x)|$ for all $x \in X$ and $0 \leqslant k \leqslant n-1$;

(5) $x \in X$, $0 \leqslant k \leqslant n$ and $f_k(x) \neq 0$ imply $\sum_{j=0}^{k} |d_j(x)| = a$.

Since $\mu(E_f(a)) < +\infty$ and $f_0(x) = 0$ for $x \notin E_f(a)$, we deduce that $f_0 \in \mathscr{L}^1$; it follows that $d_1, \ldots, d_n$, $f_1, \ldots, f_n$ belong also to $\mathscr{L}^1$. Let us also note that:

(6) If $A \in \mathscr{B}$ and $\sum_{j=0}^{n} |d_j(x)| = a$ on A, then

$$\int_A (a - |d_0(x)|)\, d\mu(x) \leqslant \int_X |f_0(x)|\, d\mu(x).$$

In fact, from (4) we deduce

$$\begin{aligned} \int_A (a - |d_0|)\, d\mu &= \sum_{j=1}^{n} \int_A |d_j|\, d\mu \leqslant \sum_{j=1}^{n} \int_X |d_j|\, d\mu \\ &= \sum_{j=1}^{n} \left(\int_X |T_j f_{j-1}|\, d\mu - \int_X |f_j|\, d\mu \right) \\ &\leqslant \sum_{j=1}^{n} \left(\int_X |f_{j-1}|\, d\mu - \int_X |f_j|\, d\mu \right) \leqslant \int_X |f_0|\, d\mu. \end{aligned}$$

Theorem 1. (Maximal ergodic theorem). – *Let* $T_0, T_1, \ldots, T_n$ *be* $n+1$ *operators belonging to* $\mathcal{D}$ *and suppose that:*

(a) $T_0 = I$;

(b) $T_{j+1} T_j = T_{j+1}$ *for each* $0 \leqslant j \leqslant n-1$.

For each $f \in \mathscr{V}$ *and each* $a > 0$ *let*

$$E_f^*(a) = \left\{ x \,\middle|\, \sup_{0 \leqslant j \leqslant n} |T_j f(x)| > a \right\}.$$

Then we have

$$a\,\mu(E_f^*(2a) \cup E_f(a)) \leqslant \int_{E_f^*(2a) \cup E_f(a)} |f|\, d\mu < +\infty.$$

In particular for $f \in \mathscr{L}^p$ *and* $a > 0$

$$\mu(E_f^*(2a)) \leqslant (N_p(f)/a)^p.$$

Proof: Let $a > 0$. By induction (using also (b)) it is easily seen that for each $1 \leqslant j \leqslant n$

$$T_j f = T_j \left(\sum_{s=0}^{j-1} d_s \right) + d_j + f_j. \tag{7}$$

Let now $x \in E_f^*(2a)$. If $|f(x)| > 2a > a$, the definition of d_0 shows that $|d_0(x)| = a$. If $|T_j f(x)| > 2a$ for some $1 \leqslant j \leqslant n$, relations (7) and (2) show that

$$2a < |T_j f(x)| \leqslant \left| T_j \left(\sum_{s=0}^{j-1} d_s \right)(x) \right| + |d_j(x)| + |f_j(x)| \leqslant a + a + |f_j(x)|;$$

hence $|f_j(x)| \neq 0$ and (use (5)) $\sum_{s=0}^{n} |d_s(x)| = a$. Thus

$$x \in E_f^*(2a) \cup E_f(a) \;\Rightarrow\; \sum_{s=0}^{n} |d_s(x)| = a.$$

By (6) above we have (use also the fact that f_0 vanishes outside $E_f(a)$):

$$\int_{E_f^*(2a) \cup E_f(a)} (a - |d_0|)\, d\mu \leqslant \int_X |f_0|\, d\mu = \int_{E_f^*(2a) \cup E_f(a)} |f_0|\, d\mu.$$

Since $|f_0|$ and $a - |d_0|$ are integrable on $E_f^*(2a) \cup E_f(a)$, it follows that $|f| - a = |f_0| + |d_0| - a$ is integrable on $E_f^*(2a) \cup E_f(a)$ and

$$\int_{E_f^*(2a) \cup E_f(a)} (|f| - a)\, d\mu \geqslant 0.$$

Since $\sum_{s=0}^{n} |d_s| \in \mathscr{L}^p$ and since

$$E_f^*(2a) \cup E_f(a) \subset \left\{ x \mid \sum_{s=0}^{n} |d_s(x)| \geqslant a \right\}$$

we deduce that $\mu(E_f^*(2a) \cup E_f(a)) < +\infty$. Therefore

$$a\mu(E_f^*(2a) \cup E_f(a)) \leqslant \int_{E_f^*(2a) \cup E_f(a)} |f|\, d\mu < +\infty.$$

Let now $f \in \mathscr{L}^p$ $(1 \leqslant p < +\infty)$ and $a > 0$. If $p = 1$ the inequality

$$\mu(E_f^*(2a)) \leqslant N_1(f)/a$$

is obvious. If $p > 1$ we apply Hölder's inequality and we obtain (here $1/p + 1/q = 1$):

$$a\mu(E_f^*(2a) \cup E_f(a)) \leqslant N_p(f)(\mu(E_f^*(2a) \cup E_f(a)))^{1/q},$$

that is,

$$\mu(E_f^*(2a) \cup E_f(a)) \leqslant (N_p(f)/a)^p.$$

This completes the proof of the theorem.

Remark. The advantage of the above method of proof is that it carries over to the abstract spaces L_E^p (E a Banach space).

Let E be a vector space and $P: E \to E$. We recall that P is a *projection* if $P \in \mathscr{L}^*(E, E)$ (that is, P is linear) and $P^2 = P$. *A sequence* (P_n) *of projections* is said to be *increasing* (respectively *decreasing*) if

$$P_n P_m = P_{\min(n,m)}$$

(respectively $P_n P_m = P_{\max(n,m)}$) for all n, m.

From Theorem 1 we deduce:

Corollary 1. – *Let* (P_n) *be an increasing (respectively decreasing) sequence of projections belonging to* $\mathscr{D}$. *Then:*

1) *For each* $f \in \mathscr{V}$, $\sup_n |P_n f(x)|$ *is finite almost everywhere.*

2) *Let* $1 \leqslant p < +\infty$ *and let* (f_k) *be a sequence in* $\mathscr{L}^p$ *such that* $\sum_k N_p(f_k) < +\infty$. *Then*

$$\limsup_{k \to +\infty} \left(\sup_n |P_n f_k(x)| \right) = 0$$

almost everywhere.

Proof: We shall only consider the case of an *increasing* sequence (P_n). (The case of a decreasing sequence can be treated similarly.)

1) Let $f \in \mathscr{L}^p \subset \mathscr{V}$ $(1 \leqslant p < +\infty)$ and let $a > 0$. For each $n \geqslant 1$ define

$$G_{f,n}^*(a) = \{x \mid \sup(|f(x)|, |P_1 f(x)|, \ldots, |P_n f(x)|) > a\};$$

define also

$$G^*_{f,\infty}(a) = \{x | \sup(|f(x)|, |P_1 f(x)|, \ldots, |P_n f(x)|, \ldots) > a\}.$$

It is clear that the sequence $(G^*_{f,n}(a))$ is increasing and that

$$\bigcup_n G^*_{f,n}(a) = G^*_{f,\infty}(a).$$

By the Maximal ergodic theorem applied to the finite sequence $T_0 = I$, $T_1 = P_n$, $T_2 = P_{n-1}, \ldots, T_n = P_1$, we have

$$\mu(G^*_{f,n}(2a)) \leqslant (N_p(f)/a)^p.$$

Passing to the limit with n we obtain:

$$\mu(G^*_{f,\infty}(2a)) \leqslant (N_p(f)/a)^p. \tag{8}$$

Since the set $\{x | \sup_n |P_n f(x)| = +\infty\}$ is contained in $G^*_{f,\infty}(b)$ for each $b > 0$, statement[1]) is proved.

2) We use here an argument due to Neveu ([108], p. 195). It suffices to show that for each $\varepsilon > 0$, the set $\limsup_k G^*_{f_k,\infty}(\varepsilon)$ is negligible. But this follows easily from inequality (8) above, since

$$\sum_k \mu(G^*_{f_k,\infty}(\varepsilon)) \leqslant \left(\frac{2}{\varepsilon}\right)^p \sum_k (N_p(f_k))^p < +\infty.$$

Hence Corollary 1 is proved.

We need now a particular case of Eberlein's mean ergodic theorem ([8], [35]). For completeness we state it in the form that is needed below:

Theorem 2. – *Let E be a reflexive Banach space. Let I be a directed set and let $(P_\alpha)_{\alpha \in I}$ be a family of elements of $\mathscr{L}(E, E)$. Suppose that:*

i) $\sup_{\alpha \in I} \|P_\alpha\| = M < +\infty$;

ii) *(P_α) is an increasing (respectively decreasing) family of projections.*

There is then $P_\infty \in \mathscr{L}(E, E)$ such that (P_α) converges strongly to P_∞. Moreover P_∞ is a projection and $P_\infty P_\alpha = P_\alpha P_\infty = P_\alpha$ for all α (respectively $P_\infty P_\alpha = P_\alpha P_\infty = P_\infty$ for all α).

Proof: We shall only consider the case of a *decreasing* family of projections. (The case of an increasing family of projections (P_α) can be reduced to the decreasing case by considering the new family (Q_α) where $Q_\alpha = I - P_\alpha$.)

---

[1]) The family $(P_\alpha)_{\alpha \in I}$ of projections is increasing (respectively decreasing) if the relations $\alpha \in I$, $\beta \in I$, $\alpha \leqslant \beta$ imply $P_\alpha P_\beta = P_\beta P_\alpha = P_\alpha$ (respectively $P_\alpha P_\beta = P_\beta P_\alpha = P_\beta$).

Let $G=\{P_\alpha|\alpha\in I\}$ and let $c(G)$ be the convex hull of G. Note that for each $T\in c(G), TP_\alpha=P_\alpha T=P_\alpha$ for all large enough α. For each $x\in G$ let

$$O(x)=\{Tx|T\in c(G)\} \qquad (=\text{the orbit of } x \text{ under } c(G))$$

and let $\overline{O(x)}$ be the closure of $O(x)$ in the norm topology. Since the set $O(x)$ is convex, $\overline{O(x)}$ coincides with the closure of $O(x)$ in the topology $\sigma(E,E')$.

We shall divide the proof into three parts:

a) We show first that for each $x\in E$ there is $y\in\overline{O(x)}$ with $Ty=y$ for all $T\in G$.

Let $x\in E$. By i) and the reflexivity of E, the family $(P_\alpha x)$ has a cluster point (=adherent value) $y\in E$ in the topology $\sigma(E,E')$. Clearly $y\in\overline{O(x)}$. It remains to show that $Ty=y$ for all $T\in G$, or equivalently that $x'(y-Ty)=0$ for all $T\in G$ and $x'\in E'$. Fix $T\in G$ and $x'\in E'$; then $z'=x'\circ T\in E'$. There is α_0 large enough that

$$TP_\alpha=P_\alpha \quad \text{for} \quad \alpha\geqslant\alpha_0. \tag{9}$$

Let now $\varepsilon>0$; there is $\beta\geqslant\alpha_0$ such that

$$|x'(y-P_\beta x)|\leqslant\frac{\varepsilon}{2} \quad \text{and} \quad |z'(P_\beta x-y)|=|x'(TP_\beta x-Ty)|\leqslant\frac{\varepsilon}{2}. \tag{10}$$

Combining (9) and (10) we obtain $|x'(y-Ty)|\leqslant\varepsilon$. Since $\varepsilon>0$ was arbitrary, this means that $x'(y-Ty)=0$ and the assertion is proved.

b) We now show that if $y\in\overline{O(x)}$ and $Ty=y$ for all $T\in G$, then $y=\lim_\alpha P_\alpha x$ (in the norm topology).

In fact, let $\varepsilon>0$. Since $y\in\overline{O(x)}$ there is $T\in c(G)$ such that

$$\|y-Tx\|\leqslant\frac{\varepsilon}{M+1}. \tag{11}$$

Now choose α' large enough that

$$P_\alpha T=P_\alpha \quad \text{for} \quad \alpha\geqslant\alpha'. \tag{12}$$

Recalling the fact that $P_\alpha y=y$ for all α and using i), (11) and (12), we deduce for $\alpha\geqslant\alpha'$

$$y-P_\alpha x=P_\alpha(y-Tx) \Rightarrow \|y-P_\alpha x\|=\|P_\alpha(y-Tx)\|\leqslant\varepsilon.$$

Hence statement b) is also proved.

c) From a) and b) it follows that for each $x\in E$, there is a *unique* element $y\in E$ with the properties $y\in\overline{O(x)}$ and $Ty=y$ for all $T\in G$; moreover $y=\lim_\alpha P_\alpha x$. *Hence* we may define $P_\infty\colon E\to E$ by

$$P_\infty x=\lim_\alpha P_\alpha x \qquad \text{for each} \qquad x\in E.$$

Clearly $P_\infty\in\mathscr{L}(E,E)$ (in fact $\|P_\infty\|\leqslant M$). Let $x\in E$ and $T\in G$. Note that $T(P_\infty x)=P_\infty x$ (from the definition) and that $P_\infty(Tx)=\lim_\alpha P_\alpha(Tx)=\lim_\alpha P_\alpha x=P_\infty x$. Thus $TP_\infty=P_\infty T=P_\infty$ for all $T\in G$. We also deduce that $P_\infty^2=P_\infty$. This completes the proof of the theorem.

Theorem 3. – *Let (P_n) be an increasing (respectively decreasing) sequence of projections belonging to $\mathscr{D}$. There is then a projection $P_\infty\in\mathscr{D}$ such that for each $f\in\mathscr{V}$, the sequence $(P_n f(x))$ converges to $P_\infty f(x)$ almost everywhere.*

Proof: We shall only consider the case of an *increasing* sequence (P_n). (The case of a decreasing sequence can be treated similarly.)

Since the space L^2 is reflexive and since $\|P_n\|_2 \leqslant 1$ for all n, we deduce from Theorem 2 that there is $P_\infty \in \mathscr{L}(L^2, L^2)$ such that the sequence (P_n) converges strongly (in L^2) to P_∞. Moreover P_∞ is a projection and $P_\infty P_n = P_n P_\infty = P_n$ for all n. Let now

$$Y = \bigcup_n P_n(L^2) \quad \text{and} \quad Z = (I - P_\infty)(L^2).$$

Then $Y+Z$ is dense in L^2 (note that for each $\tilde{u} \in L^2$, $\tilde{u} = P_\infty \tilde{u} + (I - P_\infty)\tilde{u}$). Now if $\tilde{f} \in Y$ then $\tilde{f} \in P_k(L^2)$ for some k and hence $P_n \tilde{f} = \tilde{f}$ for all $n \geqslant k$; if $\tilde{f} \in Z$ then $P_n \tilde{f} = \tilde{0}$ for all n. We deduce that $(P_n f(x))$ converges to $P_\infty f(x)$ almost everywhere, whenever $\tilde{f} \in Y+Z$.

To show that for each $\tilde{f} \in L^2$, the sequence $(P_n f(x))$ converges to $P_\infty f(x)$ almost everywhere, we use an argument due to Neveu: Let $\tilde{f} \in L^2$ and let $(\tilde{g}_k)$ be a sequence in $Y+Z$ such that

$$\sum_k N_2(f - g_k) < +\infty.$$

We have for all n and k:

$$|P_n f - P_\infty f| \leqslant \sup_m |P_m(f - g_k)| + |P_n g_k - P_\infty g_k| + |P_\infty f - P_\infty g_k|.$$

In the right hand side, for fixed k and $n \to \infty$, the second term tends to 0 almost everywhere since $\tilde{g}_k \in Y + Z$. Now if we let $k \to \infty$, the first term tends to 0 almost everywhere by statement 2) of Corollary 1, and the third term tends to 0 almost everywhere since

$$\sum_k N_2(P_\infty(f - g_k)) \leqslant \sum_k N_2(f - g_k) < +\infty.$$

It follows that $\lim_n P_n f(x) = P_\infty f(x)$ almost everywhere, as asserted. In particular we see that the restriction of P_∞ to $L^1 \cap L^\infty$ is a Dunford-Schwartz operator (use Fatou's lemma).

Finally $L^1 \cap L^\infty$ is contained and dense in every L^p, $1 \leqslant p < +\infty$. Repeating the argument of Neveu (given above for L^2) we deduce that $(P_n f(x))$ converges to $P_\infty f(x)$ almost everywhere, for every $f \in \mathscr{V}$. This concludes the proof of Theorem 3.

Appendix II. Notation and terminology

1. – For each set X we denote by $\mathscr{P}(X)$ the set of all parts of X. If $A \in \mathscr{P}(X)$ we denote by φ_A the mapping of X into R defined by

$$\varphi_A(x) = \begin{cases} 1 & \text{if } x \in A, \\ 0 & \text{if } x \notin A. \end{cases}$$

We call φ_A the characteristic function of A.

If X is a set, $Y \subset X$ and f a function with domain X, we denote by $f|Y$ the *restriction* of f to Y.

Let X be a set. A set $\mathscr{T} \subset \mathscr{P}(X)$ is called a *tribe* (or a σ-algebra) of parts of X if:

1.1) $X \in \mathscr{T}$;

1.2) $A \in \mathscr{T}$ implies $\complement A \in \mathscr{T}$;

1.3) If (A_n) is a sequence of sets belonging to $\mathscr{T}$ then $\bigcup_n A_n \in \mathscr{T}$.

Note that a tribe $\mathscr{T}$ is a non-void set since by 1.1), $\mathscr{T} \ni X$; by 1.2), $\mathscr{T} \ni \varnothing$. If X is a set then $\mathscr{P}(X)$ is a tribe. If $(\mathscr{T}_j)_{j \in J}$ is any family of tribes of parts of X then $\bigcap_{j \in J} \mathscr{T}_j$ is also a tribe. Hence given $\mathscr{E} \subset \mathscr{P}(X)$ there exists a smallest tribe, denoted $\mathscr{T}(\mathscr{E})$, which contains $\mathscr{E}$. We say that $\mathscr{T}(\mathscr{E})$ is the tribe *spanned* by $\mathscr{E}$.

Let X be a topological space. The tribe spanned by the *closed* parts of X is called the *Borel tribe* of X. It is clear that the Borel tribe of X is the tribe spanned by the *open* parts of X. A set $A \subset X$ is called a *Borel set* if A belongs to the Borel tribe of X.

Let X be a set and $\mathscr{T}$ a tribe of parts of X. A function $f\colon X \to \bar{R}$ is said to be *measurable with respect to* $\mathscr{T}$ or $\mathscr{T}$*-measurable* if

1.4) $$f^{-1}(B) \in \mathscr{T}$$

for *every* Borel set $B \subset \bar{R}$. It is easy to see that $f\colon X \to \bar{R}$ is measurable with respect to $\mathscr{T}$ if

1.5) $$\{x \mid f(x) < a\} \in \mathscr{T}$$

for *every* $a \in R$.

A set $A \subset X$ belongs to $\mathscr{T}$ if and only if φ_A is measurable with respect to $\mathscr{T}$.

Let X be a *topological space.* A mapping $f: X \to \bar{R}$ is said to be *Borel measurable* if f is measurable with respect to the Borel tribe of X. If $f: X \to \bar{R}$ is continuous then f is Borel measurable.

Let X be a set and $\mathscr{T}$ a tribe of parts of X. The set of all mappings of X *into* R, measurable with respect to $\mathscr{T}$, is an algebra with respect to the usual laws of composition. Moreover if $f_1, \ldots, f_p$ are *mappings* of X into $\bar{R}$ measurable with respect to $\mathscr{T}$ and if $\varphi: \bar{R}^p \to \bar{R}$ is Borel measurable then $\varphi(f_1, \ldots, f_p)$ is measurable with respect to $\mathscr{T}$.

Let (f_n) be a sequence of mappings of X into $\bar{R}$, which converges pointwise to a mapping f; if each f_n is measurable with respect to $\mathscr{T}$, then f is measurable with respect to $\mathscr{T}$.

Let $\mathscr{S}(\mathscr{T})$ be the set of all functions of the form

1.6) $$\sum_j \lambda_j \varphi_{A_j} \quad \text{(finite sum)}$$

where $\lambda_j \in R$ and $A_j \in \mathscr{T}$. Note that in 1.6) we may suppose $A_{j'} \cap A_{j''} = \emptyset$ if $j' \neq j''$. The functions belonging to $\mathscr{S}(\mathscr{T})$ are called simple functions "based" on sets belonging to $\mathscr{T}$. It is easy to see that $\mathscr{S}(\mathscr{T})$ is a subalgebra of the algebra of all mappings of X into R, measurable with respect to $\mathscr{T}$.

If $f: X \to \bar{R}$ is positive and measurable with respect to $\mathscr{T}$, then there exists an increasing sequence of positive functions belonging to $\mathscr{S}(\mathscr{T})$, which converges uniformly to f with respect to the uniform structure of $\bar{R}$. If $f: X \to R$ is *bounded* and measurable with respect to $\mathscr{T}$, then there exists a sequence of functions belonging to $\mathscr{S}(\mathscr{T})$, which converges uniformly to f with respect to the usual uniform structure of R.

2. – Let X be a set. A mapping $\Gamma: \mathscr{P}(X) \to \bar{R}_+$ is an *outer measure in the sense of Carathéodory* (or Carathéodory outer measure) if

2.1) $\Gamma(\emptyset) = 0$;

2.2) $\Gamma(A) \leqslant \Gamma(B)$ if $A \subset B$;

2.3 If (A_n) is a sequence of parts of X then

$$\Gamma\left(\bigcup_n A_n\right) \leqslant \sum_n \Gamma(A_n).$$

A set $A \subset X$ is *Γ-measurable* if

2.4) $$\Gamma(Y) = \Gamma(Y \cap A) + \Gamma(Y \cap \complement A)$$

for all $Y \in \mathscr{P}(X)$. The set $\mathscr{T}(\Gamma)$ of all Γ-measurable sets is a tribe; moreover the restriction of Γ to $\mathscr{T}(\Gamma)$ is a *countably additive* mapping of $\mathscr{T}(\Gamma)$ into $\bar{R}_+$. If $A \in \mathscr{T}(\Gamma)$, $B \subset A$ and $\Gamma(A) = 0$ then $B \in \mathscr{T}(\Gamma)$.

A function $f: X \to \bar{R}$ is called Γ-measurable if it is measurable with respect to $\mathscr{T}(\Gamma)$.

3. – A vector space E over R endowed with an order relation "$\leqslant$" is called an *ordered vector space* if:

3.1) $x \leqslant y \Rightarrow x+z \leqslant y+z$ for all $z \in E$;

3.2) $x \geqslant 0$ and $\lambda \geqslant 0 \Rightarrow \lambda x \geqslant 0$.

A *Riesz space* is an ordered vector space E such that

$$\sup(x,y) \quad \text{and} \quad \inf(x,y)$$

exist for every x and y in E.

If X is a set and $\mathscr{T}$ a tribe of parts of X then the set of all mappings of X into R, measurable with respect to $\mathscr{T}$, endowed with the usual structure of a vector space and the usual order relation is a Riesz space. If X is a topological space then the set $C_R(X)$ of all continuous functions on X to R, endowed with the usual structure of a vector space and the usual order relation is a Riesz space[1]). The set $C_R^b(X)$, of all *bounded* continuous functions on X to R, endowed with the usual structure of a vector space and the usual order relation is a Riesz space.

4. – For every two vector spaces E and F we denote by $\mathscr{L}^*(E,F)$ the set of all linear mappings of E into F endowed with the usual structure of a vector space. If $F=R$ (and E is a vector space over R) we write E^* instead of $\mathscr{L}^*(E,R)$; we call E^* the *algebraic* dual of E.

For $x \in E$ and $x^* \in E^*$ we often write

$$\langle x, x^* \rangle \quad \text{instead of} \quad x^*(x).$$

Let now E and F be *normed spaces*. We denote by $\mathscr{L}(E,F)$ the set of all *continuous* linear mappings of E into F; clearly

$$\mathscr{L}(E,F) \subset \mathscr{L}^*(E,F).$$

We also write E' instead of $\mathscr{L}(E,R)$; we call E' the *dual* of E (or the topological dual).

If we write

$$\|u\| = \sup_{\|x\| \leqslant 1} \|u(x)\|$$

for all $u \in \mathscr{L}(E,F)$ then $u \to \|u\|$ is a norm on $\mathscr{L}(E,F)$. If F is a Banach space, then $\mathscr{L}(E,F)$ endowed with this norm is a *Banach space*. The dual E' of a normed space E is always a Banach space (whether or not E is a Banach space).

A Banach space F is of *countable type* if there exists a countable set F_0 dense in F.

---

[1]) The smallest tribe of parts of X with respect to which all functions $f \in C_R(X)$ are measurable is called the *Baire tribe* of X. A set $A \subset X$ is called a *Baire set* if A belongs to the Baire tribe of X. A mapping $f: X \to \bar{R}$ is said to be *Baire measurable* if f is measurable with respect to the Baire tribe of X.

Open problems

Let X be a locally compact space and μ a positive Radon measure on X with $\operatorname{Supp}\mu = X$. If X is metrizable, then there is a strong lifting of $M_R^\infty(X,\mu)$ (see Theorem 3, section 2, chapter 8). There are also many examples of non-metrizable spaces X such that a strong lifting of $M_R^\infty(X,\mu)$ exists (see chapter 8). The following problem is not yet solved:

(A) *Decide whether or not there exists a strong lifting of* $M_R^\infty(X,\mu)$, *for every* X *and* μ.

As it is shown in chapter 9, problem (A) is in a certain sense equivalent with the problem of disintegration of measures. Another equivalent formulation of (A) is given in Theorem 2, section 5, chapter 10.

Suppose that $(X,N,\mathcal{R})$ is strictly localizable. Let $\mathcal{A} \subset M_R^\infty(X,N,\mathcal{R})$ be a subalgebra of $M_R^\infty(X,N,\mathcal{R})$ with the property that if f, g belong to $\mathcal{A}$ and $f \equiv g$ then $f = g$.

Consider now the following problem:

(B) *Find conditions on* $(X,N,\mathcal{R})$ *and* $\mathcal{A}$ *which imply the existence of a lifting* ρ *of* $M_R^\infty(X,N,\mathcal{R})$ *such that* $\rho(f) = f$ *for each* $f \in \mathcal{A}$.

Problem (B) is of course related to problem (A) and was in fact suggested by the notion of strong lifting. In connection with problem (B) see also Theorem 6 in section 3, chapter 8.

Suppose that $(X,N,\mathcal{R})$ is strictly localizable. Let $\mathcal{A} \subset M_R^\infty(X,N,\mathcal{R})$ be a subalgebra of $M_R^\infty(X,N,\mathcal{R})$ with the property that for each $f \in M_R^\infty(X,N,\mathcal{R})$ there is $f' \in \mathcal{A}$ such that $f' \equiv f$.

Consider now the following problem:

(C) *Find conditions on* $(X,N,\mathcal{R})$ *and* $\mathcal{A}$ *which imply the existence of a lifting* ρ *of* $M_R^\infty(X,N,\mathcal{R})$ *such that* $\rho(f) \in \mathcal{A}$ *for each* $f \in M_R^\infty(X,N,\mathcal{R})$.

Suppose in particular that X is a locally compact Polish space, μ a positive Radon measure on X and $\mathcal{A}$ the subalgebra of $M_R^\infty(X,\mu)$ consisting of all Borel measurable functions. We recall that in this case a lifting ρ of $M_R^\infty(X,\mu)$ is called a *Borel lifting* if $\rho(f) \in \mathcal{A}$ for each

$f \in M_R^\infty(X,\mu)$ (see the Appendix at the end of chapter 8). The question of the existence of a Borel lifting of $M_R^\infty(X,\mu)$ [respectively, the existence of a strong Borel lifting of $M_R^\infty(X,\mu)$, when $\operatorname{Supp}\mu = X$] is only partially solved (namely using the continuum hypothesis, in [105]).

Suppose that $(X,N,\mathscr{R})$ is strictly localizable. Denote by $\mathscr{H}(X,N,\mathscr{R})$ the group of all bijections s: $X \to X$ having the following two properties:

1) $f \circ s$ and $f \circ s^{-1}$ belong to $M_R^\infty(X,N,\mathscr{R})$ if f to $M_R^\infty(X,N,\mathscr{R})$;

2) $f \circ s$ and $f \circ s^{-1}$ are N-negligible if f is N-negligible.

Note that for each $s \in \mathscr{H}(X,N,\mathscr{R})$ the mapping $f \to f \circ s$ is an isomorphism of $M_R^\infty(X,N,\mathscr{R})$ onto itself; clearly $f \equiv g$ implies $f \circ s \equiv g \circ s$. Let now $\mathscr{G} \subset \mathscr{H}(X,N,\mathscr{R})$. A linear lifting or a lifting ρ of $M_R^\infty(X,N,\mathscr{R})$ *commutes* with $\mathscr{G}$ if:

$$\rho(f \circ s) = \rho(f) \circ s,$$

for all $f \in M_R^\infty(X,N,\mathscr{R})$ and $s \in \mathscr{G}$.

Consider now the following problem:

(D) *Find conditions on* $(X,N,\mathscr{R})$ *and* $\mathscr{G}$ *which imply the existence of a linear lifting or a lifting of* $M_R^\infty(X,N,\mathscr{R})$ *commuting with* $\mathscr{G}$.

For relevant contributions to problem (D) the reader may consult [54] and[72].

Bibliography

[1] ALFSEN, E. M.: Some covering theorems of Vitali type. Math. Ann. **159**, 203—216 (1965).

[2] ANDO, T.: Contractive projections in L_p-spaces. Pacific J. Math. **17**, 391—405 (1966).

[3] ARENS, R.: Representation of *-algebras. Duke math. J. **14**, 269—282 (1947).

[4] AUMANN, G.: Reelle Funktionen. Berlin-Göttingen-Heidelberg: Springer 1954.

[5] BADE, W. G.: The space of all continuous functions on a compact Hausdorff space, Section 6, Mimeographed Notes. Berkeley: University of California 1957.

[6] BAHADUR, R. R.: Measurable subspaces and subalgebras. Proc. Amer. Math. Soc. **6**, 565—570 (1955).

[7] BANACH, S.: Théorème sur les ensembles de première catégorie. Fund. Math. **16**, 395—398 (1930).

[8] BARRY, J. Y.: On the convergence of ordered sets of projections. Proc. Amer. Math. Soc. **5**, 313—314 (1954).

[9] BAUER, H.: Šilovscher Rand und Dirichletsches Problem. Ann. Inst. Fourier **11**, 89—136 (1961).

[10] BECK, A., H. H. CORSON, and A. B. SIMON: The interior points of the product of two subsets of a locally compact group. Proc. Amer. Math. Soc. **9**, 648—652 (1958).

[11] BIRKHOFF, GARRETT: Lattice theory. New York: A.M.S. 1948.

[12] — — Moyennes des fonctions bornées, algèbre et théorie des nombres, Colloques Internationaux du C.N.R.S., no. **24**, pp. 143—153. Paris: Centre National de la Recherche Scientifique 1950.

[13] BONSALL, F. F., J. LINDENSTRAUSS, and R. R. PHELPS: Extreme positive operators on algebras of functions. Math. Scand. **18**, 161—182 (1966).

[14] BOURBAKI, N.: Topologie générale, Chap. 1—10. Paris: Hermann 1958—1963.

[15] BOURBAKI, N.: Intégration, Chap. 1—8. Paris: Hermann 1959—1967.

[16] BOURBAKI, N.: Espaces vectoriels topologiques, Chap. 1—5. Paris: Hermann 1953—1955.

[17] BRUNK, H. D.: On an extension of the concept of conditional expectation. Proc. Amer. Math. Soc. **14**, 298—304 (1963).

[18] CARTIER, P., J. M. G. FELL, and P. A. MEYER: Comparaison des mesures portées par un ensemble convexe compact. Bull. Soc. Math. France **92**, 435—445 (1964).

[19] COMFORT, W., and H. GORDON: Vitali's theorem for invariant measures. Trans. Amer. Math. Soc. **99**, 83—90 (1961).

[20] DAY, M. M.: Fixed-point theorems for compact convex sets. Illinois J. Math. **5**, 585—590 (1961).

[21] DIEUDONNÉ, J.: Sur le théorème de Lebesgue-Nikodym, III. Ann. Univ. Grenoble **23**, 25—53 (1948).

[22] — Sur le théorème des Lebesgue-Nikodym, IV. J. Indian Math. Soc., N.S. **15**, 77—86 (1951).

[23] DINCULEANU, N.: Sur la représentation intégrale des certaines opérations linéaires II. Compositio Math. **14**, 1—22 (1959).

[24] —, and C. FOIAŞ: Mesures vectorielles et opérations linéaires sur L_E^p. Comptes Rendus **248**, 1759—1762 (1959).

[25] — — Sur la représentation intégrale de certaines opérations linéaires IV. Canad. J. Math. **13**, 529—556 (1961).

[26] — Integral representation of vector measures and linear operations. Studia Math. **25**, 181—205 (1965).

[27] — Vector measures. Berlin: Deutscher Verlag der Wissenschaften 1966.

[28] DIXMIER, J.: Les algèbres d'opérateurs dans l'espace hilbertien (Algèbres de von Neumann). Paris: Gauthier-Villars 1957.

[29] DOLÉANS, CATHERINE: Désintégration des mesures, Séminaire Choquet (Initiation à l'analyse). Paris: 1964.

[30] DONOGHUE, W. F.: On the lifting property. Proc. Amer. Math. Soc. **16**, 913—914 (1965).

[31] DOOB, J. L.: Stochastic processes. New York: Wiley 1953.

[32] DOUGLAS, R. G.: Contractive projections on an $\mathscr{L}_1$ space. Pacific J. Math. **15**, 443—462 (1965).

[33] DUNFORD, N., and J. B. PETTIS: Linear operations on summable functions. Trans. Amer. Math. Soc. **47**, 323—392 (1940).

[34] —, and J. T. SCHWARTZ: Linear operators, Part I. New York: Interscience 1958

[35] EBERLEIN, W. F.: Abstract ergodic theorems and weak almost periodic functions. Trans. Amer. Math. Soc. **67**, 217—240 (1949).

[36] EDWARDS, R. E., and E. HEWITT: Pointwise limits for sequences of convolution operators. Acta Math. **113**, 181-218 (1965).

[37] ELLIS, A. J.: Extreme positive operators. Oxford Quarterly J. Math. **15**, 342-344 (1964).

[38] FILLMORE, P. A.: On topology induced by measure. Proc. Amer. Math. Soc. **17**, 854-857 (1966).

[39] FOGUEL, S. R., and L. TZAFRIRI: Remarks on contractive projections in L_p-spaces, to appear.

[40] FOIAS, C., and I. SINGER: Some remarks on the representation of linear operators in spaces of vector-valued continous functions. Revue Math. Pures et Appl. **5**, 729-752 (1960).

[41] FOMIN, S. V.: On measures invariant under certain groups of transformations. Izv. Akad. Nauk. SSSR. Ser. Mat. **14**, 261-274 (1950).

[42] GOFFMAN, C., and D. WATERMAN: Approximately continuous transformations. Proc. Amer. Math. Soc. **12**, 116-121 (1961).

[43] —, C. NEUGEBAUER, and T. NISHIURA: Density topology and approximate continuity. Duke Math. J. **28**, 497—506 (1961).

[44] GOULD, G. G., and M. MAHOWALD: Measures on completely regular spaces. J. London Math. Soc. **37**, 103—111 (1962).

[45] GROTHENDIECK, A.: Critères de compacité dans les espaces fonctionnels généraux. Amer. J. Math. **74**, 168—186 (1952).

[46] — Produits tensoriels topologiques et espaces nucléaires. Memoirs Amer. Math. Soc. **16** (1955).

[47] — Espaces vectoriels topologiques. São Paulo: 1958.
[48] Halmos, P. R., and J. von Neumann: Operator methods in classical mechanics, II. Ann. of Math. **43,** 332—350 (1942).
[49] — On a theorem of Dieudonné. Proc. Nat. Acad. Sci., USA **35,** 38—32 (1949).
[50] — Measure theory. New York: Van Nostrand 1950.
[51] Haupt, O., and C. Pauc: La topologie approximative de Denjoy envisagée comme vraie topologie. C. R. Acad. Sci., Paris **234,** 390—392 (1952).
[52] — — Über die durch allgemeine Ableitungsbasen bestimmten Topologien. Ann. Mat. Pura Appl. Ser. IV **36,** 247—271 (1954).
[53] Hille, E., and R. S. Phillips: Functional analysis and semi-groups. Providence: A.M.S. 1957.
[54] Ionescu Tulcea, A.: On the lifting property (V). Ann. Math. Stat. **36,** 819—828 (1965).
[55] — Sur le relèvement fort et la désintégration des mesures. Comptes Rendus Acad. Sci., Paris **262,** 617—618 (1966).
[56] — Sur la domination des mesures et la désintégration des mesures. Comptes Rendus Acad. Sci., Paris **262,** 1142—1445 (1966).
[57] — Liftings compatible with topologies. Bull. Soc. Math. de Grèce **8,** 116—126 (1967).
[58] — On the lifting property, Proceedings Symposium in Analysis, Queen's University, Kingston Ontario (June, 1967).
[59] Ionescu Tulcea, C.: Deux théorèmes concernant certains espaces de champs de vecteurs. Bull. Sci. Math. **79,** 106—111 (1955).
[60] — On the lifting property and disintegration of measures, invited address presented before the Chicago Meeting of the A.M.S. on April 9, 1965. Bull. Amer. Math. Soc. **71,** 829—842 (1965).
[61] — Remarks on the lifting property and the disintegration of measures. Technical report, U.S. Army Research Office, Durham N.C. (June 1965).
[62] — Sur certains endomorphismes de $L_C^\infty(Z,\mu)$. Comptes Rendus Acad. Sci., Paris **261,** 4961—4963 (1965).
[62'] — Two theorems concerning the disintegration of measures, J. Math. Anal. Appl., to appear.
[63] Ionescu Tulcea, A., and C. Ionescu Tulcea: A remark concerning extremal points, mimeographed note (1960).
[64] — — On the decomposition and integral representation of continuous linear operators. Annali di Matem. Pura ed Applicata **53,** 63—87 (1961).
[65] — — On the lifting property, I. J. Math. Anal. Appl. **3,** 537—546 (1961).
[66] — — Abstract ergodic theorems. Proc. Nat. Acad. Sci. U.S.A. **48,** 204—206 (1962).
[67] — — On the lifting property. II, Representation of linear operators on spaces $L_E^r, 1 \leq r < \infty$. J. Math. Mech. **11,** 773—796 (1962).
[68] — — Abstract ergodic theorems. Trans. Amer. Math. Soc. **107,** 107—124 (1963).
[69] — — Problems and remarks concerning the lifting property. Technical report, U.S. Army Research Office, Durham, N.C., September 1963.
[70] — — On the lifting property, III. Bull. Amer. Math. Soc. **70,** 193—197 (1964).
[71] — — On the lifting property. IV, Disintegration of measures. Ann. Inst. Fourier (Grenoble) **14,** 445—472 (1964).
[72] — — On the existence of a lifting commuting with the left translations of an arbitrary locally compact group, Proceedings Fifth Berkeley Symposium on Math. Stat. and Probability p. 63—97. Univ. of California Press 1967.

[73] — — Liftings for abstract valued functions and separable stochastic processes, to appear.

[74] JACOBS, K.: Ergodic theory. I, II. Matematisk Institut Aarhus Universitet Aarhus: 1963.

[75] JERISON, M. and G. RABSON: Convergence theorems obtained from induced homomorphisms of a group algebra. Ann. of Math. **63**, 176—190 (1956).

[76] JIRINA, MILOSLAV: On regular conditional probabilities. Czechoslovak Math. J. **9**, 445—450 (1959).

[77] KAKUTANI, S.: Concrete representations of abstract (*L*)-spaces and the mean ergodic theorem. Annals Math. **42**, 523—537 (1941).

[78] KÖLZOW, D.: Differentiation von Massen. Lecture Notes in Mathematics **65**, Berlin-Heidelberg-New York: Springer 1968.

[79] KRICKEBERG, K., and C. PAUC: Martingales et dérivation. Bull. Soc. Math. France **91**, 455—544 (1963).

[80] LECAM, L.: Convergence in distribution of stochastic processes. Univ. Calif. Publ. Statist. **2**, 207—236 (1957).

[81] LLOYD, S. P.: On extreme averaging operators. Proc. Amer. Math. Soc. **14** 305—310 (1963).

[82] LOÈVE, M.: Probability theory. New York: Van Nostrand, 1963.

[83] LOOMIS, L. H.: Unique direct integral decompositions on convex sets. Amer. J. Math. **84**, 509—526 (1962).

[84] LUXEMBURG, W. A. J.: Banach Function Spaces. Thesis Delft Inst. of Techn., Assen (Netherlands) 1955.

[85] —, and A. C. ZAANEN: Notes on Banach Function Spaces, Proc. Acad. Sci., Amsterdam; Note I, A66, 135—147 (1963); Note II, A66, 148—153 (1963); Note III, A66, 239—250 (1963); Note IV, A66, 215—263 (1963); Note V, A66, 496—504 (1963); Note VI, A66, 655—668 (1963); Note VII, A66, 669—681 (1963); Note VIII, A67, 104—119 (1964); Note IX, A67, 360—376 (1964); Note X, A67, 493—506 (1964); Note XI, A67, 507—518 (1964); Note XII, A67, 519—529 (1964); Note XIII, A67, 530—543 (1964).

[86] MCSHANE, E. J.: Images of sets satisfying the condition of Baire. Ann. of Math. **51**, 380—386 (1950).

[87] MAHARAM, D.: On homogeneous measure algebras. Proc. Nat. Acad. Sci. U.S.A. **28**, 108—111 (1942).

[88] — On a theorem of von Neumann. Proc. Amer. Math. Soc. **9**, 987—994 (1958).

[89] — On two theorems of Jessen. Prob. Amer. Math. Soc. **9**, 995—999 (1958).

[90] — Automorphismus of products of measure spaces. Proc. Amer. Math. Soc. **9**, 702—707 (1958).

[91] MARTIN, N. F. G.: A topology for certain measure spaces. Trans. Amer. Math. Soc. **112**, 1—18 (1964).

[92] MARTINEAU, A.: Sur le théorème du graphe fermé. Comptes Rendus Acad. Sci., Paris **263**, 870—871 (1966).

[93] MÉTIVIER, M.: Sur la désintégration des mesures. C. R. Acad. Sci., Paris **256**, 1062—1965.

[94] MÉTIVIER, M.: Limites projectives de mesures. Martingales, applications, Thèse (Faculté des Sciences, Université de Rennes), Juin 1963.

[95] MEYER, P. A.: Progrès recent dans la théorie des cônes convexes à base compacte. Séminaire Brelot-Choquet-Deny (Théorie du potentiel), Paris, 1963/64.

[96] — Probability and potentials. Blaisdell Publ. Comp. 1966.

[97] MOKOBODZKI, G.: Barycentres généralisés, Séminaire Brelot-Chocquet-Deny (Théorie du potentiel). Paris, 1961/62.

[98] MONTGOMERY, D., and L. Zippin: Topological transformation groups. New York: Interscience 1955.

[99] MOY, S.-T.: Characterizations of conditional expectation as a transformation on functions spaces. Pacific J. Math. **4,** 47—63 (1954).

[100] MUELLER, B. J.: Three results for locally compact groups connected with the Haar measure density theorem. Proc. Amer. Math. Soc. **16** 1414—1416 (1965).

[101] NAIMARK, M. A.: Normed rings. Moscow 1956.

[102] VON NEUMANN, J.: Algebraische Repräsentanten der Funktionen bis auf eine Menge von Masse Null. J. Crelle **165,** 109—115 (1931).

[103] — Einige Sätze über meßbare Abbildungen. Ann. of Math. **33,** 574—586 (1932).

[104] — Zur Operatorenmethode in der klassischen Mechanik. Ann. of Math. **33,** 587—642 (1932).

[105] —, and M. H. STONE: The determination of representative elements in the residual classes of a Boolean algebra. Fund. Math. **25,** 353—378 (1935).

[106] NEVEU, J.: Deux remarques sur la théorie des martingales. Zeit. Wahrscheinlichkeitstheorie **3,** 122—127 (1964).

[107] — Relations entre la théorie des martingales et la théorie ergodique, Colloque International du C.N.R.S. sur la Théorie du Potentiel, Summer 1964.

[108] — Bases mathématiques du calcul des probabilités. Paris: Masson 1964.

[109] ONICESCU, O., G. MIHOC, and C. T. IONESCU TULCEA: Calculus of Probability (in Roumanian). Roumanian Academy 1956.

[109′] OXTOBY, J. C.: Spaces that admit a category measure. J. Reine Angew. Math. **205,** 156—170 (1960).

[110] PANZONE, R., and C. SEGOVIA: Measurable transformations on compact spaces and o.n. systems on compact groups. Rev. Un. Mat. Argentina **22,** 83—102 (1964).

[111] PETTIS, B. J.: On continuity and openness of homomorphisms in topological groups. Ann. of Math. **52,** 293—308 (1950).

[112] — Remarks on a theorem of E. J. McShane. Proc. Amer. Math. Soc. **2,** 166—171 (1951).

[113] PHELPS, R. R.: Extremal operators and homomorphisms. Trans. Amer. Math. Soc. **108,** 265—274 (1963).

[114] PHILLIPS, R. S.: On linear transformations. Trans. Amer. Math. Soc. **48,** 516—541 (1940).

[115] — On weakly compact subsets of a Banach space. Amer. Math. J. **65,** 108—136 (1943).

[116] RINGROSE, J. R.: On well-bounded operators II. Proc. London Math. Soc. **13,** 613—638 (1963).

[117] RISS, JEAN: Les semi-normes dénombrablement convexes. Publ. Sci. Univ. Alger, Ser. A **3,** 107—120 (1956).

[118] ROBERTSON, A. P., and WENDY ROBERTSON: On the space of subsets of a uniform space, Proc. Amer. Math. Soc. **12,** 321—326 (1961).

[119] RYAN, R.: The lifting property and direct sums, MRC, Technical Report No. **328,** U.S. Army. University of Wisconsin 1962.

[120] RYAN, R.: Representative sets and direct sums. Proc. Amer. Math. Soc. **15,** 387—390 (1964).

[121] SAKS, S.: Theory of the integral. New York: Hafner 1937.

[122] SCHWARTZ, L.: Lectures on the theory of Radon measures on arbitrary topological spaces. Bombay: Tata Institute, forthcoming

[123] — Sur le théorème du graphe fermé. Comptes Rendus Acad. Sci. Paris **263**, 870—871 (1966).

[124] — Désintégration régulière d'une mesure par rapport à une famille de tribus. Comptes Rendus Acad. Sci. Paris **266**, 424—425 (1968).

[125] — Applications des désintégrations régulières. Comptes Rendus Acad. Sci., Paris **266**, 467—469 (1968).

[126] SEGAL, I. E.: Equivalences of measure spaces. Amer. J. Math. **73**, 275—313 (1951).

[127] SINGER, I.: Sur les applications linéaires majorées des espaces de fonctions continues. Atti Accad. Naz. Lincei Rend. **27**, 35—41 (1959).

[128] SINGER, I.: Sur la représentation intégrale des applications linéaires continues des espaces L^p $(1 \leq p < \infty)$. Atti Accad. Naz. Lincei Rend. **29** 28—32 (1960).

[129] STONE, M. A.: Application of the theory of Boolean rings to general topology. Trans. Amer. Math. Soc. **41**, 375—481 (1937).

[130] STONE, M. H.: Notes on integration. Proc. Nat. Acad. Sci. U.S.A.; Note I **34**, 336—342 (1948); Note II **34**, 447—455 (1948); Note III **34**, 483—490 1948); Note IV **35**, 50—58 (1949).

[131] STØRMER, E.: Positive linear maps of operator algebras. Acta Math. **110**, 233—278 (1963).

[132] STRASSEN, V.: The existence of probability measures with given marginals. Ann. Math. Stat. **36**, 423—439 (1965).

[133] TROYER, R., and W. ZIEMER: Topologies generated by outer measures. J. Math. Mech. **12**, 485—494 (1963).

[134] VESTERSTRØM, J., and W. WILS: On point realizations of L^∞-endomorphisms. Aarhus Matematisk Institut, Aarhus Universitet 1968.

[135] WRIGHT, F.: Generalized means. Trans. Amer. Math. Soc. **98**, 187—203 (1961); ibid. **100**, 370 (1961).

[136] ZAANEN, A. C.: The Radon-Nikodym theorem, I, II. Indag. Math. **23**, 157—187 (1961).

[137] ZYGMUND, A.: Trigonometric series, Vol. II. Cambridge Univ. Press 1959.

Subject Index

List of Symbols

Ergebnisse der Mathematik und ihrer Grenzgebiete

---

1. Bachmann: Transfinite Zahlen. DM 38,—; US $ 9.50
2. Miranda: Partial Differential Equations of Elliptic Type. DM 58,—; US $ 14.50
4. Samuel: Méthodes d'Algèbre Abstraite en Géométrie Algébrique. DM 26,—; US $ 6.50
5. Dieudonné: La Géométrie des Groupes Classiques. DM 38,—; US $ 9.50
6. Roth: Algebraic Threefolds with Special Regard to Problems of Rationality. DM 19,80; US $ 4.95
7. Ostmann: Additive Zahlentheorie. 1. Teil: Allgemeine Untersuchungen. DM 38,—; US $ 9.50
8. Wittich: Neuere Untersuchungen über eindeutige analytische Funktionen. DM 28,–; US $ 7.00
11. Ostmann: Additive Zahlentheorie. 2. Teil: Spezielle Zahlenmengen. DM 22,—; US $ 5.50
14. Coxeter/Moser: Generators and Relations for Discrete Groups. DM 32,—; US $ 8.00
16. Cesari: Asymptotic Behavior and Stability Problems in Ordinary Differential Equations. DM 36,—; US $ 9.00
17. Severi: Il teorema di Riemann-Roch per curve. Superficie e varietà questioni collegate. DM 23,60; US $ 5.90
18. Jenkins: Univalent Functions and Conformal Mapping. DM 34,—; US $ 8.50
19. Boas/Buck: Polynomial Expansions of Analytic Functions. DM 16,—; US $ 4.00
20. Bruck: A Survey of Binary Systems. DM 36,—; US $ 9.00
21. Day: Normed Linear Spaces. DM 17,80; US $ 4.45
23. Bergman: Integral Operators in the Theory of Linear Partial Differential Equations. DM 36,—; US $ 9.00
25. Sikorski: Boolean Algebras. DM 38,—; US $ 9.50
26. Künzi: Quasikonforme Abbildungen. DM 39,—; US $ 9.75
27. Schatten: Norm Ideals of Completely Continuous Operators. DM 23,60; US $ 5.90
28. Noshiro: Cluster Sets. DM 36,—; US $ 9.00
29. Jacobs: Neuere Methoden und Ergebnisse der Ergodentheorie. DM 49,80; US $ 12.45
30. Beckenbach/Bellman: Inequalities. DM 30,—; US $ 7.50
31. Wolfowitz: Coding Theorems of Information Theory. DM 27,—; US $ 6.75
32. Constantinescu/Cornea: Ideale Ränder Riemannscher Flächen. DM 68,—; US $ 17.00
33. Conner/Floyd: Differentiable Periodic Maps. DM 26,—; US $ 6.50
34. Mumford: Geometric Invariant Theory. DM 22,—; US $ 5.50
35. Gabriel/Zisman: Calculus of Fractions and Homotopy Theory. DM 38,—; US $ 9.50
36. Putnam: Commutation Properties of Hilbert Space Operators and Related Topics. DM 28,—; US $ 7.00
37. Neumann: Varieties of Groups. DM 46,—; US $ 11.50
38. Boas: Integrability Theorems for Trigonometric Transforms. DM 18,—; US $ 4.50
39. Sz.-Nagy: Spektraldarstellung linearer Transformationen des Hilbertschen Raumes. DM 18,—; US $ 4.50

40. Seligman: Modular Lie Algebras. DM 39,—; US $ 9.75
41. Deuring: Algebren. DM 24,—; US $ 6.00
42. Schütte: Vollständige Systeme modaler und intuitionistischer Logik. DM 24,—; US $ 6.00
43. Smullyan: First-Order Logic. DM 36,—; US $ 9.00
44. Dembowski: Finite Geometries. DM 68,—; US $ 17.00
45. Linnik: Ergodic Properties of Algebraic Fields. DM 44,—; US $ 11.00
46. Krull: Idealtheorie. DM 28,—; US $ 7.00
47. Nachbin: Topology on Spaces of Holomorphic Mappings. DM 18,—; US $ 4.50
48. Ionescu Tulcea/Ionescu Tulcea: Topics in the Theory of Lifting. DM 36,—; US $ 9.00
49. Hayes/Pauc: Derivation and Martingales. In preparation